Artificial Habitats for Marine and Freshwater Fisheries

ARTIFICIAL HABITATS FOR MARINE AND FRESHWATER FISHERIES

EDITED BY

William Seaman, Jr.

Florida Sea Grant College Program, and
Department of Fisheries and Aquaculture
University of Florida
Gainesville, Florida

Lucian M. Sprague

Agricultural and Rural Development Department
World Bank
Washington, D.C. (retired), and
Department of Fisheries and Aquaculture
University of Florida
Gainesville, Florida

ACADEMIC PRESS, INC.
Harcourt Brace Jovanovich, Publishers
San Diego   New York   Boston   London   Sydney   Tokyo   Toronto

Copyright © 1991 by ACADEMIC PRESS, INC.
All Rights Reserved.
No part of this publication may be reproduced or transmitted in any form or by any means, electronic or mechanical, including photocopy, recording, or any information storage and retrieval system, without permission in writing from the publisher.

Academic Press, Inc.
San Diego, California 92101

United Kingdom Edition published by
Academic Press Limited
24–28 Oval Road, London NW1 7DX

Library of Congress Cataloging-in-Publication Data

Artificial habitats for marine and freshwater fisheries / [edited
  by] William Seaman, Jr., and Lucian M. Sprague.
      p.   cm.
    Includes index.
    ISBN 0-12-634345-4
    1. Fish habitat improvement.   I. Seaman, William, Jr., date.
  II. Sprague, Lucian M.
  SH157.8.A78   1991
  639.3--dc20   91-14729
                                                               CIP

PRINTED IN THE UNITED STATES OF AMERICA
91 92 93 94   9 8 7 6 5 4 3 2 1

Let There Be a
Firmament
in the Midst of the
Waters
Genesis 1:6

Contents

Contributors

Numbers in parentheses indicate the pages on which the authors' contributions begin.

R. F. AMBROSE (61), Marine Science Institute, University of California, Santa Barbara, Santa Barbara, California 93106

J. A. BOHNSACK (61), Southeast Fisheries Center, National Marine Fisheries Service, Miami, Florida 33149

S. A. BORTONE (177), Department of Biology, University of West Florida, Pensacola, Florida 32504

R. S. GROVE (109), Advanced Engineering, Southern California Edison Co., Rosemead, California 91770

D. L. JOHNSON (61), School of Natural Resources, Ohio State University, Columbus, Ohio 43210

J. J. KIMMEL (177), Florida Department of Natural Resources, St. Petersburg, Florida 33701

J. M. McGURRIN (31), Atlantic States Marine Fisheries Commission, Washington, D.C. 20036

J. W. MILON (237), Department of Food and Resource Economics, University of Florida, Gainesville, Florida 32611

M. NAKAMURA (109), Tokyo University of Fisheries, Tokyo, Japan

J. J. POLOVINA (154), Southwest Fisheries Center, Honolulu Laboratory, National Marine Fisheries Service, Honolulu, Hawaii 96822-2396

W. SEAMAN, JR. (1, 31), Florida Sea Grant College Program, and Department of Fisheries and Aquaculture, University of Florida, Gainesville, Florida 32611

C. J. SONU (109), Tekmarine, Inc., Pasadena, California 91101

L. M. SPRAGUE (1, 31), Agricultural and Rural Development Department, World Bank, Washington, D.C.,[1] and Department

---

1. Retired from World Bank.

of Fisheries and Aquaculture, University of Florida, Gainesville, Florida 32611

R. B. STONE (31), Recreational and Interjurisdictional Fisheries, National Marine Fisheries Service, Silver Spring, Maryland 20910

Preface

The ancients who observed that fishes were more abundant in proximity to floating debris not only increased their catches, but also by their insight and ingenuity set the stage for an important field in fishery science. Today, the creation of artificial habitats in freshwater and marine environments is commonplace, reflecting global efforts to deploy natural and man-made materials and structures. As in other scientific areas dealing with living natural resources, a primary intent has been to increase harvests of plants and animals.

Interest in artificial habitats and the technologies related to altering aquatic ecosystems is high in many areas of the world and developing in others. It includes both organizations and individuals involved in artisanal, commercial, and recreational fishing, as well as those interested in fisheries management and aquatic environmental preservation and mitigation.

From the earliest aquatic artificial habitat technologies to the present, most practical applications have been introduced by fishermen. Only recently have scientists begun to coalesce, interpret, and expand a body of technical knowledge for this broad subject. Fishery managers have a longer history of involvement, but much of their experience is anecdotal. Reference to the study of artificial habitats in most textbooks of fishery science either is nonexistent or brief and restricted to specific techniques and situations. Historically, this reflects both the scattered nature of the technical information and the paucity of quantitative data about artificial habitats. By contrast, journalists have written numerous articles for the lay reader owing to the popularity and intrinsic appeal of this subject.

In the past decade, however, the situation has changed markedly. For example, a critical mass of technical information emerged in concert with the Third and Fourth International Conferences on Artificial Habitats for Fisheries held in Newport Beach, California, and Miami, Florida in 1983 and 1987, respectively. In Florida, attendees from twenty-six nations exchanged a significant number of new findings concerning habitats and associated recreational, commercial, and small-scale (traditional) fisheries, as well as environmental issues in this field. It also became clear that the increase in scientific research (principally as independent studies to document what many observers often referred to as an art form) had not been adequately accompanied by comparative analysis that resulted in a coherent and organized synthesis of information.

As an initial effort to foster such an evaluation, this volume has three purposes: First, to fill the need for a review of the subject by presenting facts and issues that are emerging in the worldwide utilization of artificial habitats in aquatic ecosystems; second is to present advances in scientific investigations, including ecology, engineering, socioeconomics, and monitoring and assessment; third is to make the material accessible to a wider audience by bringing the study of artificial habitats into the mainstream of fisheries science and management. The rather broad audience includes those who deploy habitats and harvest the resource, those who set policy for natural resources, and the academic, scientific, and engineering community.

Typically, most current technical information on artificial habitats is contained in specialists journals or nonreviewed "gray" literature that present the results of diverse individual research projects and management situations. This volume seeks to address the subject from as broad a geographic, disciplinary, and time perspective as possible given the formative, fragmented, and diverse nature of the available material.

When we began this book, our intent was to provide as much information about global trends in the field as possible. In practice we have not entirely achieved our goal because the worldwide literature in this area of fisheries is more limited and scattered than we believed initially. Nevertheless, the editors and authors have kept the goal in sight. For instance, although the review of history includes all inhabited continents, lack of written technical documentation for newer programs shifts some of the focus to more extensively described experiences in Japan and the United States. The coverage of ecology, by its nature global, is good, although until recently the data base has been based principally on descriptive and survey research that provides a basis for more experimental and theoretical approaches.

Chapter Four focuses almost exclusively on larger fabricated structures and high-technology practices and thus the Japanese experience. Fortunately, this creates a unique opportunity to make Japan's work accessible to an English language audience. Meanwhile the smaller scale, traditional, and less technologically complex experiences of others are reflected in several places in the text. In the discussion of fisheries impact on artificial habitats, evidence is presented concerning interactions of fishery populations and assemblages, fishing effort and yield, and habitat in a variety of settings. Finally, the evaluation of habitats is presented from two perspectives: the socioeconomic perspective, which has a remarkably short history for this subject, and the environmental monitoring and assessment perspective, which has developed a wide spectrum of techniques.

We believe that great opportunities exist to use artificial habitats (1) in classical scientific investigations of ecosystem structure and function (e.g., ecology, animal behavior, dynamics of nutrients and food webs); (2) as a basis for engineering advances in the field of underwater technology; and (3) as tools in fishery and environmental management since, with carefully planned additional research and development, there is an increasingly important role for artificial habitats in multidisciplinary approaches in these areas. We expect that science and policy will catch up with rapidly developing applications in this field.

Acknowledgments

Chapters were peer-reviewed individually by selected authorities in the field, as follows: H. Ansley and S. Shipman, Georgia Department of Natural Resources; M. R. Barnett, University of Florida; E. E. DeMartini and S. G. Pooley, U.S. National Marine Fisheries Service; P. E. Gadd, Coastal Frontiers Corporation (United States); M. L. Harmelin-Vivien, Centre d'Oceanologie de Marseille, Université D'Aix-Marseille (France); C. Itosu, Tokyo University of Fisheries; A. R. Longhurst, Bedford Institute of Oceanography (Canada); J. N. Suhayda, Louisiana State University; and D. R. Talhelm, Michigan State University. Technical comments on the overall volume were provided by J. C. Marr. We appreciate the extensive review provided by these colleagues.

More limited comments and additional information on artificial habitat practices were provided by V. L. Aprieto, University of the Philippines in the Visayas; G. D. Ardizzone, Universita di Roma, Italy; E. Baquiero, Instituto Nacional de la Pesca, Mexico; K. L. Branden, South Australia Department of Fisheries; K. Collins, Southampton University, England; E. deBernardi, Association Monegasque pour la Protection de la Nature, Monaco; A. Kocatas, Ege University, Turkey; A. A. Konan, Centre de Recherches Oceanographique, Ivory Coast; M. Gomez Llorente, Centro de Technologia Pesquere, Canary Islands; C. K. Looi, Malaysia Department of Fisheries; A. Markevich, Academy of Sciences, Union of Soviet Socialist Republics; A. D. McIntyre, University of Aberdeen, Scotland; A. Naeem, Ministry of Fisheries and Agriculture, Maldives; G. L. Preston, South Pacific Commission; G. Relini, Universita di Genova, Italy; P. J. Sanjeeva Raj, Center for Research on New International Order, India; K. Shao, Academia Sinica, Republic of China; E. Spanier, University of Haifa, Israel; and Soon Kil Yi, Korea Ocean Research and Development Institute, to whom we are grateful. We also thank W. R. Gordon, Jr., Southwest Texas State University, and R. E. Lange, New York State Department of Environmental Conservation, for earlier comments on the manuscript.

Each contributor expresses thanks for the assistance of their employers and office staff, and J. Bohnsack acknowledges the following for his chapter:

B. Bohnsack for helpful criticisms and suggestions; the National Marine Fisheries Service and the Florida Sea Grant College Program (Grant No. R/ LR-B-22) for partial funding of research reported; National Oceanic and Atmospheric Administration, U.S. Department of Commerce; the Rosenstiel School of Marine and Atmospheric Science, University of Miami; Dade County Department of Environmental Resources Management; U.S. Precast Corporation; and Byrd Commercial Diving.

Support for preparation of this volume includes funding from Exxon Corporation, for which we especially thank K. C. Williams. We gratefully acknowledge the role of C. Wilson of Louisiana State University in facilitating this sponsorship. Also, funding from the International Conference on Artificial Habitats for Fisheries, and office support from the Florida Sea Grant College Program, headquartered at the University of Florida (UF) and supported in part by Grant NA89AA-D-SG053 from the U.S. Department of Commerce National Oceanic and Atmospheric Administration, made this work possible. J. Cato, UF, encouraged all phases of this project. We thank P. Rolfe, UF, for skillfully typing portions of the document and making final revisions. Much of the manuscript was typed by M. Little, whose long-term involvement in our research and writing is deeply appreciated. Numerous illustrations in the text were prepared by J. Wallenka of the UF Office of Instructional Resources. K. Hale of the marine library of the University of Miami provided timely access to recent bibliographic citations.

Authors in this book commented on their colleagues' manuscripts at different stages of preparation. We thank them for the insights they provided in that process and, of course, also for their enduring commitment to the collegial process of writing this book.

Finally, our special thanks go to our families in Woodbridge, Virginia and Gainesville, Florida for their support when we worked at each other's home, when we worked away from home, and all the times in between. Individually, L. Sprague wishes to gratefully acknowledge the early and key support for his work by Mary and Helen McAndrews, John E. Cushing, Garrett Hardin, Clyde Stormont, and John C. Marr, all of whom gave far more than they knew, when it counted most. W. Seaman acknowledges the wisdom and special role of C. R. Gilbert, H. L. Popenoe, and the late E. C. Raney in the events leading to production of this book.

1

Artificial Habitat Practices in Aquatic Systems

W. SEAMAN, JR.
Florida Sea Grant College Program, and
Department of Fisheries and Aquaculture
University of Florida
Gainesville, Florida

L. M. SPRAGUE[1]
Agricultural and Rural
Development Department
World Bank
Washington, D.C., and
Department of Fisheries and Aquaculture
University of Florida
Gainesville, Florida

I. Introduction

The magnitude of the world's fishery harvest from bodies of fresh and salt-water makes it a subject of active interest globally. The quality of life for peoples in waterfront villages, the economy of regions dependent on commercial fishing, and the aesthetics of recreation are intertwined with the productivity of aquatic ecosystems. In numerous local areas and increasingly at national levels, these interests have enhanced the production and harvest of plant and animal species through remarkably successful modifications of the aquatic environment. Most applications of artificial habitat technologies have occurred in the last few decades.

Globally, organizations and individuals have deployed a rather diverse array of structures to influence behavior and ecology of aquatic organisms. Effective habitat-enhancement practices range from small-scale modifications of local environments to complex engineering of large structures deployed over extensive areas of seafloor. Having arisen principally by chance

1. Retired from World Bank.

Artificial Habitats for Marine and Freshwater Fisheries
Copyright © 1991 by Academic Press, Inc.
All rights of reproduction in any form reserved.

1

and employed historically as an art, this field and even its oldest materials, structures, and practices are now the subject of accelerating scientific inquiry. Despite success such as establishing a major tuna fishery in the tropical Pacific, providing numerous recreational fishing opportunities in North America, developing significant seafood resources in coastal Asia, or growing shellfish in southern Europe, fundamental questions about biological processes and economic efficiency remain and are of practical consequence to investors and policy-makers.

This book addresses achievements, issues, and concerns in the field of artificial habitat enhancement, defined as the manipulation of natural aquatic habitats through the addition of man-made or natural structure, principally to enhance fisheries but also to influence the life cycle of organisms or the function of ecological systems for other purposes. A growing body of technical knowledge has emerged around these practices in marine and freshwater environments. From beginnings based on anecdotal observations and purely descriptive studies, it has come to include experimental results as well. The applicable principles are consistent with allied branches of science, such as forestry, wildlife management, agriculture, and aquaculture.

The premise of this book is that this developing field now represents a distinct multidisciplinary branch of fishery science. It includes ecology, engineering, economics, geography, sociology, law, and other disciplines, which separately and together build a significant global body of experience and understanding. Artificial habitat enhancement is at a point in its development where linkage among disciplines is clearly to be encouraged, particularly in view of significant analytical documentation in the last decade.

II. Diversity and Definitions of Habitats

It is reasonable to speculate that perceptive Neolithic observations at different shorelines of the world doubtless led to independent efforts to deploy materials that fostered a greater harvest of organisms. Much later, the first technical literature chronicled the 18th century establishment of fisheries in coastal Asia and later in North America, in conjunction with the use of submerged objects. The traditional rationale for introducing structures, specifically to increase the abundance of organisms of interest to fisheries, has been expanded to include general enhancement of environmental quality. More recently, structure as a means of disposing wastes has emerged as a questionable and controversial rationale.

Collectively, artificial habitat development practices now are significant in artisanal, commercial, and recreational fishery settings as well as in environmental management. Some of the oldest practices are used in artisanal situations, that is, in small-scale, so-called traditional fisheries in which the population of a settlement or village might subsist on the harvest from nearby habitats, including those that have been modified by the addition of natural materials. More recently, other interests employ greater levels of technology to fabricate and deploy structures. Representative materials and structural types are depicted in Fig. 1.1 and are discussed below. Despite the diverse interests and geographic locations involved, a small number of basic materials, structures, and deployment strategies existed historically. They have been augmented by innovations only in recent years. Chapter 2 provides a synopsis of documented events.

As recently as 1981 the textbook by Everhart and Youngs exhorted fishery professionals to extend their concepts about habitat improvement beyond the prevailing simple designs dating from the 1930s. (See, e.g., Fig. 1.2.) In the decade since that text, not only biologists but also ocean-ographers, engineers, economists, planners, managers, and policy-makers have made great strides in extending concepts for the manipulation of habi-tat, illustrating the rapidity with which the field has evolved. Much of the evolution has been in the marine sector, as reviewed in the following chap-ters. Freshwater advances are reviewed also, particularly for ecological un-derstanding (see Chapter 3). Some of the traditional and proven practices, specifically those smaller habitat structures used in streams or lakes, which have been described thoroughly in books and journal articles (e.g., Everhart and Youngs, 1981; Brown, 1986), are not reviewed in as much detail here. Typical structures of this kind include log deflectors, stone-filled cribs, and brushpiles.

Usually, freshwater structures have been relatively small and deployed as individual units or in a small array, so that their influence is localized. By contrast, the artificial habitats emphasized in this book are physically larger (either singly or as modules that are grouped) and influence a greater area biologically. Typically they have been used most widely in coastal and ocean settings, and only to a limited degree in lakes and reservoirs. Thus, the engineering aspects of handling these structures may be formidable and their economic and social impact significant.

The structures in Fig. 1.1 represent a spectrum of approaches that (1) are characterized variously as "low" or "high" technology, (2) require different types of manufactured or natural materials in construction, and (3) differ in terms of expense for materials, labor, and other resources. As discussed later in this section and in Chapter 2, extreme approaches can range from principally volunteer efforts that deploy cheap, surplus materials to subsidized private and public partnerships that use carefully fabricated structures.

Historically, the most common classifications of artificial habitats reflect physical location in the water column, composition and construction of

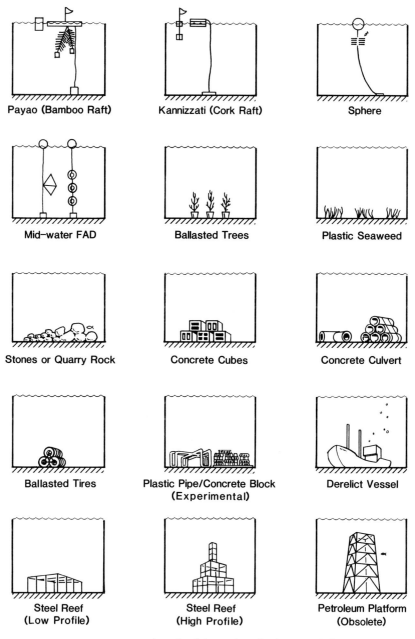

Figure 1.1 Representative examples of widely used artificial aquatic habitat materials and structures.

Figure 1.2 The log deflector is a traditional and accepted structure used extensively in freshwater fishery management, along with other devices to increase habitat diversity.

materials, and nature of the environment where deployed. Thus, habitats might be described as a "benthic concrete pipe reef," or a "floating foam-filled tire," or a "lake brushpile." While habitats also might be defined in terms of the management objectives identified later in this chapter, it is still uncommon to have a biological definition that is specific to functions such as reproduction, nursery, or forage sites for individual species or life history stages. No glossary of terms has been published for this subject.

Among the oldest habitat enhancement practices is the use of floating structures made of natural materials to attract finfishes. In the Philippines and the northern Mediterranean Sea, respectively, inexpensive surface rafts of bamboo and cork have been employed for centuries. But it was only in the early 1960s that scientists at the U.S. Bureau of Commercial Fisheries Biological Laboratory in Hawaii began systematic efforts to understand the mechanisms underlying the attractive effect of floating structures (Manar, 1966). This phenomenon has been addressed by others because of its importance to the surface fishery for tuna (Shomura, 1977; Stequert and Marsac, 1989). Improvements in the design of anchoring systems for floating structures used in the Pacific Islands were reviewed by Boy and Smith (1984). Recent experiments with other materials such as steel and fiberglass floating spheres of various sizes have been conducted in the western Pacific and Atlantic Oceans. Much of the effort has been in coastal artisanal fisheries, as reviewed by the U.S. National Research Council (1988).

Floating structures, commonly referred to as fish-aggregating devices (FADs), are used in freshwater as well as in the ocean. FADs are deployed at the surface or various levels in the water column, and they may be made of automobile tires or synthetic mesh which is suspended in the water column. Typical midwater FADs are depicted in Fig. 1.1.

Benthic structures typically are smaller in freshwater systems than on the bottom of the sea. In lakes and reservoirs, trees, bundles of brush, plastic "grass," bound tires (all appropriately weighted), and piles of stones are used in various situations. (See Phillips, 1990, for numerous drawings.) Such structures also have been used in marine settings, ordinarily in estuarine or protected waters. Larger and denser materials are used in the open ocean. It is these latter habitat structures that are commonly referred to as "artificial reefs." This term is used frequently in the following chapters, but the broader term "artifical habitat" is commended to the reader as a means of including structures such as artificial seaweed. In turn, a broader view of these practices is encouraged.

Typical marine artificial-reef habitats—intended to mimic natural reefs at least by providing relief on flat, featureless ocean bottoms—include concrete or steel modules, frames, and other structures made to certain design specifications; natural products such as quarry rock; and scrap man-made items including concrete culvert or other building materials (e.g., bridge demolition rubble); and steel structures such as storage tanks, petroleum production platforms, and ships. The latter class of items often are acquired and deployed opportunistically, and referred to as "materials of opportunity." The challenges posed by design, handling, and placement of massive seafloor structures are reviewed in Chapter 4.

The recurrent use globally of basic FAD and reef designs may indicate that technological activity in the development of new designs has reached a plateau. However, growth of activity to deploy the structures like those illustrated in this chapter and throughout this book has been explosive. Nowhere is the deployment of reefs more heavily practiced than in Japan, where nearly 10% of the coastal seafloor has been enhanced by a network of carefully designed concrete, steel, and other structures (Japan National Coastal Fishery Development Association, 1987). Whether science has kept pace with the growing body of practitioners around the world (described in Chapter 2), is a question we shall explore in examining the ecological and engineering performance of artificial habitats (Chapters 3 and 4), their impacts on fisheries (Chapter 5), techniques for evaluating their environmental characteristics (Chapter 6), and economic impacts (Chapter 7).

III. Worldwide Utilization of Artificial Habitats

Artificial habitats, particularly of a large-scale, have been deployed in fresh-water and especially marine settings around the world, although descriptions of them are not consistently recorded in published technical literature. Major areas of activity include the basins of the Caribbean and Mediterranean Seas, Southeast Asia, Japan, Australia, North America, and some South Pacific Ocean islands according to recent accounts and summaries identified here and more extensively in the next chapter. Lesser levels of activity are reported for Central America, northern and central Europe, and limited areas of Africa and South America. Table 1.1 presents a qualitative summary of the relative levels of application of these technologies worldwide, according to three principal aquatic interests. This information is based primarily on a limited number of available literature descriptions. Absence of recorded activity in certain zones may reflect a lack of either published information, documented observations, or effort. Documentation is an issue with which future observers need to be concerned.

The relative levels of activity devoted to artificial habitats vary significantly among geographic regions and among the three principal sectors that

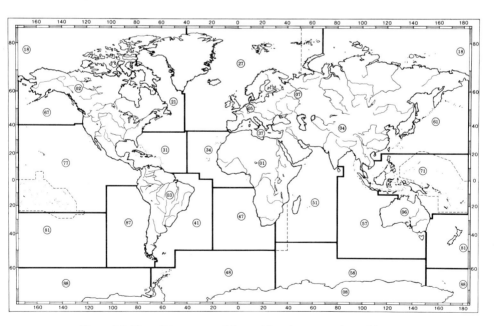

Figure 1.3 Major fishing areas of the world as characterized in Table 1.1.

TABLE 1.1

Occurrence of Artificial Habitats in Major Fishing Areas of the World, Principally in Inland and Coastal Oceanic Waters

World region, area and FAO code[b]	Level of Activity[a]		
	Artisanal fishing	Commercial fishing	Recreational fishing or diving
Asia[c]			
Inland (04)	+		
Pacific, Northwest (61)	+ +	+ + +	+
Pacific, W. Central (71)	+ + +	+ + +	+
Indian Ocean, Eastern (57)	+	+ +	+
Oceania (Australia-New Zealand region)[c]			
Inland (06)			
Pacific, W. Central (71) (includes S. Pacific Islands)	+	+	+ +
Pacific, Southwest (81)			+
Indian Ocean, Eastern (57)			+
Africa[c]			
Inland (01)	+ +		
Indian Ocean, Western (51)	+		
Atlantic, Southeast (47)			
Atlantic, E. Central (34)	+	+	
Mediterranean (37)	+	+	
Europe[c]			
Inland (05, 07)		+	
Mediterranean (37)	+ +	+ +	+
Atlantic, Northeast (27)		+	
North America[c]			
Inland (02)		+	+ + +
Atlantic, Northwest (21)			+ +
Atlantic, W. Central (31)	+	+ +	+ + +
Pacific, E. Central (77) (includes Hawaii)	+	+ +	+
Pacific, Northwest (67)			+
South America (continental)[c]			
Inland (03)			
Atlantic, W. Central (31)	+	+	+
Atlantic, Southwest (41)			
Pacific, Southwest (87)			
Pacific, E. Central (77)			

Areas and code numbers conform to usage of the Department of Fisheries, United Nations Food and Agricultural Organization.

[a] Activity levels: + + +, highest; + +, moderate; +, some; blank space, none reported according to literature or knowledgeable observers.

[b] As an exception to our adherence to using published technical reports as the basis for this volume, here we also draw from the observations of knowledgeable persons.

[c] Numbers refer to regions marked on map. (See Fig. 1.3 on previous page.)

promote and utilize them. Artisanal fishing, which relies on low-cost, easily made structures to enhance subsistence harvests for local family and village members, is centered in coastal Asia, with secondary areas of effort along the northern Mediterranean Sea and inland coastal Africa. Artisanal efforts are particularly prominent in the Philippines, Thailand, Indonesia, Malaysia, Maldive Islands, India, and islands of the South Pacific Ocean, frequently using modules of natural materials. (See e.g., Figs. 1.4, 5.1, 5.2 and 5.3.) Commercial fishing uses more powerful equipment to harvest and sell fishes taken from FADs and reefs most commonly located in coastal Asia, the eastern Indian Ocean, the Caribbean and northern Mediterranean basins, and the islands of the South Pacific. Japan, the Philippines, Taiwan, Italy, and Monaco are notable locations, with emphasis on high density benthic structures or floating raft technologies (Figs. 1.5 and 5.4). Principal areas for recreational fishing and some diving include coastal and inland

Figure 1.4 Some of the highest fishery yields taken at artificial habitats come from the payao, a floating bamboo raft used in pelagic fisheries of the Philippines. Both artisanal and commercial fisheries operate here.

Figure 1.5 Benthic artificial habitats of steel and other durable materials are manufactured in factories and deployed with heavy construction equipment for use in commercial fishing in Japan's coastal ocean. (Photograph courtesy of the Kozai Club, Tokyo.)

North America, Australia, and some islands of the South Pacific. This sector has placed the greatest reliance on surplus industrial and construction materials and sunken ships. Recreational diving, including some development of support industry to serve tourists, is a more recent pursuit in these areas, particularly in the United States (See Fig. 1.6 and 1.7.) Finally, an emerging application of artificial habitat technologies is for environmental mitigation, an endeavor most prominently developed along the temperate Pacific coast of the United States as a means of ameliorating habitat destroyed by shoreline development (e.g., Hueckel *et al.*, 1989; see also Chapter 5).

As might be expected, the harvest of species from this global network of habitats is diverse and reflects the resident pool of plants and animals. Due to incomplete reporting, it is not possible to compile exhaustive landings data. The limited literature that summarizes regional trends indicates that significant harvests are based on, among other groups: tuna in the Philippines, South Pacific and Hawaiian islands, and Indian Ocean (U.S. National Research Council, 1988; Buckley *et al.*, 1989); various seaweeds, abalone, and marine finfishes such as jacks, mackerels, rockfishes, flounder, and sea bream in Japan (Mottet, 1981); groupers and snappers in the Gulf of Thailand (Pramokchutima and Vadhanakul, 1987); tilapia and other finfishes in Africa (Welcomme, 1972); mussels, lobster, jacks, mullets, seabass, and

Figure 1.6 Materials of opportunity, used in many artificial habitat projects for recreational fishing enhancement include surplus or scrap construction products such as concrete culvert and derelict vessels.

Figure 1.7 Surplus or derelict vessels are used not only as artificial habitats for fisheries but also as a dive site for a growing recreational scuba diving audience. Particular emphasis may be in clear waters that lack natural relief or to alleviate heavy usage of stressed natural systems such as the coral reefs. (Photograph courtesy of Monroe County Tourist Development Council, Florida.)

mackerels along the seacoast of southern Europe (Bombace, 1989); shellfish in the Mediterranean Sea (General Fisheries Council for the Mediterranean, 1986); spiny lobsters in the western Caribbean basin (U.S. National Research Council, 1988); and sunfishes and basses in freshwater and various jacks, seabasses, and groupers in the nearshore oceans of North America (Mc-

Gurrin *et al.*, 1989). Habitat programs in these and other areas are described in the next chapter, and fishery aspects are reviewed in Chapter 5.

IV. Planning

Of all practices in fishery science, the use of artificial habitat is one of the most ecumenical. Whereas in some areas complete control of an artificial habitats program may fall to a single agency or ministry, elsewhere multiple interests may be involved in planning, building or using the habitats. This process can involve, sometimes simultaneously, public agencies, businesses, private nonprofit organizations, scientists, managers, users of the resource, and the lay public. Planning in either context, whether for an entire program or a single project, can be complex. In this section we identify some of the goals of such interests and elements of planning needed to achieve them.

Only in the last decade have planning documents begun to appear. Most are focused on local or regional areas, and tend to circulate only to readers in the immediate area affected. In some places, practices of individuals and groups to plan, construct and use habitats simply have evolved over time.

Artificial habitats are increasingly popular in the aquatic environment because, responsibly used, they can either enhance the success or efficiency of exploitation of a fishery resource or else facilitate achievement of other fishery or environmental management objectives. Commonly, the goal of habitat deployment is stated very generally, such as to provide or "improve" vertical relief of bottom habitat, so that in turn the biomass of a fishery population increases locally. The uses and purposes of artificial habitats were identified by Bohnsack and Sutherland (1985) to include the following: (1) improvement of fishing; (2) reduction of fishing on defined stocks; (3) restriction of fishing from shipping lanes; (4) prevention of trawling in defined areas, as a means of resolving disputes over fishing grounds; (5) mitigation of detrimental environmental impacts and restoration of habitat; (6) control of beach erosion; (7) provision of breakwaters; and (8) provision of spawning areas. Recent additional objectives are (9) provision of recreational diving sites; (10) creation of areas for scientific experiments; and (11) setting up "farms" for bivalve molluscs with the primary purpose of recycling nutrients. In addition, structures not intended as a habitat may nevertheless provide it as a secondary function (e.g., offshore platforms, docks).

A. Perspectives on Planning

Ideally, the addition of man-made structure to an aquatic system would be consistent with its ecology, while management goals would be consistent with environmental constraints and user interests. To date, no generic guidebook or handbook of procedures worldwide for comprehensively planning artificial habitat has been published. However, some national-level documents certainly contain some generic information, as do publications focused on certain user interests or specific logistic steps (e.g., Stone, 1985; Japan Coastal Fisheries Promotion Association, 1986; and others noted below).

In discussing a planning framework for organizing habitat enhancement efforts, our assumption is that no hidden agenda for creation of the habitat exists and that planning is perceived as a desirable first step in the creation of the habitat (Meier et al., 1989). Unfortunately, lesser motives may have played an important role in the establishment of some artificial habitats, developed for the primary purpose of disposing of industrial or municipal scrap and solid waste (e.g., tires, car bodies, used concrete) or as a publicity tool.

It is an issue of considerable concern in some quarters that various groups may lack basic information about the criteria for chemical composition and physical structure of proposed habitat materials and may assume, therefore, that any reasonably stable material may be safely deposited in the aquatic environment as a disposal technique that will incidentally benefit fishery interests. Furthermore, the potential for failure of artificial habitat structures must be understood clearly. Although there may be many anecdotal accounts of unsuccessful or failed habitats, lack of technical documentation is a weakness of the database.

It appears to us that, except for Japan and the United States, a national plan or program for aquatic artificial habitat development is atypical. By plan we mean an objective statement of goals for artificial habitat utilization, in the context of national fishery, economic, and environmental policies, accompanied by guidance on ecological, engineering and other logistic constraints, and information on biogeographic variation over large areas. More commonly, habitats are built when local governments or interest groups independently initiate deployment efforts. Sometimes a national fishery agency will initiate a local project. While such habitat deployments may conform to regulatory requirements in a given political jurisdiction, over a small ecological region, typically plans are not well articulated because they fail to clearly specify goals and objectives as part of a comprehensive national strategy. To our knowledge, comparative analyses or reviews of habitat planning at national or even local levels are absent from the established literature, and only limited examination of specific situations has appeared. A working paper by Murray (1989) discusses policy issues in the southeastern United States, and its recommendations have in turn been taken up by the

Figure 1.8 Sequence of planning steps for conducting aquatic artificial habitat projects.

U.S. Atlantic States Marine Fisheries Commission (Reeff *et al.*, 1990). (The reader is directed to case studies of national planning in Chapter 2.)

Particularly for projects at the local or regional level, a desirable sequence of events from preliminary planning to postdeployment evaluation of artificial habitat is depicted in Fig. 1.8. Ideally one or more goals are established at the outset of the process and various assessments of effectiveness are performed after the habitat is deployed. Artificial habitat projects should begin with consideration and emphasis of social needs and assessment of social and economic costs and benefits; proceed to identification of siting and construction alternatives; and finally, carry out monitoring and assessment of biological and socioeconomic impacts. Key data and archival information should become part of a consistent computerized database system. A remarkably diverse set of disciplines, interests, and considerations should enter into any decision regarding the deployment of long-lived artificial habitats. Among the pertinent disciplines are ecology, engineering, economics, sociology, and resource planning. It is important that they interact. A series of planning phases particularly relevant to individual projects is presented by Ditton and Burke (1985).

For purposes of this discussion, three broad categories of decisions apply to the planning process: institutional, engineering, and ecological (Fig. 1.9). Institutional aspects include setting goals, assessing needs of and impacts on fisheries and other interests, and complying with or establishing a regulatory framework. Engineering aspects include assessment of physical environmental factors significant to the habitat; establishing design criteria; selection, fabrication or assembly, and deployment of materials; and monitoring performance of the habitat. Finally, ecological aspects include the determination and possibly prediction of overall environmental impacts of the habitat, and specifically how it contributes to or affects populations or assemblages of particular organisms.

Evaluation of performance is an important consideration for each category. In practice, limited resources often preclude an appropriate emphasis on evaluation. Lack of funding may restrict evaluation in formal programs, whereas the independent nature of local, informal, and individual projects

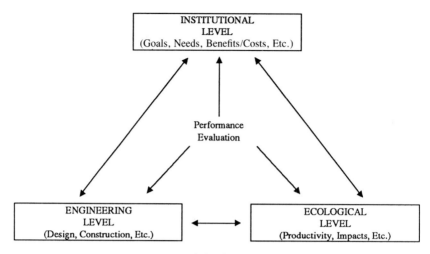

Figure 1.9 Framework for planning artificial habitats.

may not even recognize the need to assess performance. (The paucity of literature is discussed in Chapters 3 and 7.)

It is useful to consider the categories of decisions concerning artificial habitat development from a hierarchical perspective. That is, initially an institutional phase is necessary, first to determine objectives and feasibility, and then to initiate a process that considers logistic and evaluation requirements. Logically, the principal government agency with authority for fishery or habitat planning would be responsible. As a minimum, a thoughtful—and interdisciplinary—review of the possible issues to be addressed before the deployment of habitat would appear to be essential. Both users of the resource and scientists responsible for design and evaluation should participate early in the process. Written documentation even at this stage is important if a body of experience is to be assembled for guiding future efforts.

Once a decision has been made at some level of authority or responsibility to proceed with either a comprehensive program or an individual project to develop an artificial habitat, then engineering considerations become paramount. Again, the integration of disciplines is an aspect that needs to be stressed, for example in matching the biological requirements of a given species with the physical properties of habitat materials and their configuration. From an ecological point of view, it is commonplace to carry out biological or predeployment surveys of potential sites in an attempt to avoid conflict with important existing habitats (e.g., coral reefs). However, once a man-made structure is deployed, the work of ecological characterization of a site often does not continue with a detailed assessment of the biological community structure. (The lack of literature reports attests to this.) This is regrettable because a failure to adequately monitor biological performance inevitably results in an inability to evaluate the degree to which the habitat meets its original objectives. Concern for this process has been expressed by Bohnsack and Sutherland (1985). Meanwhile, economic and physical performance must be conducted as well. The subjects of monitoring, assessment, and evaluation are the basis for Chapters 6 and 7. (Readers wishing to explore the subject of planning in greater depth are referred to the works of Agarwala, 1982; Falundi, 1973; Friedman, 1987; and Rittel and Webber, 1973.) Finally, it is important to realize that different constraints may be important depending on whether the setting is national, regional, or local.

V. Research Perspectives for Manipulating Habitat

Scientists involved in efforts to manipulate aquatic habitat through the use of artificial structure are united by the common goal of addressing ways to overcome limiting factors in order to alter the productivity of the system. This involves bringing scientific methods to bear on the environmental constraints. Thus a body of systematized knowledge has grown up in concert with advances in practice. It is not surprising, however, that this growth has been accompanied by opposing viewpoints, as indicated below.

While the Earth's major environments vary greatly in the degree to which they have been studied, it is helpful to our perspective of artificial aquatic habitats to recognize that agriculture, forestry, fisheries, range, and wildlife management share some commonality of principles and practices. For example, in addition to concerns for understanding ecological-limiting factors, these sciences also must reconcile basic objectives of enhancing biological yields with the political and economic realities of multiple-use concepts for natural resources. Meanwhile, they differ greatly in the time span of their application, sophistication of method, and degree of scientific understanding. It is beyond the scope of this book to analyze the literature, history, and development of these companion bodies of knowledge, but clearly they should offer a perspective for aquatic habitat enhancement research, with reference to past development, present complexity, and future scope (see, e.g., Yoakum et al., 1980). We also commend the approach of drawing information from the basic studies of natural systems from botany, geology, zoology, microbiology, ecology, and oceanography to understand and predict artificial habitats. The blurring of distinctions between natural and artificial

Figure 1.10 Artificial habitat programs increasingly use fabricated materials deployed in pre-determined patterns to influence biotic diversity. Underwater scenes depict biota that resemble natural reef systems. Sequence of photographs depicts (**A**) concrete blocks at a staging area on shore; (**B**) coastal deployment; and (**C,D**) colonization by organisms. (Photographs courtesy of Association Monegasque pour la Protection de la Nature [Monaco].)

habitats, at least from the perspective of physical appearance of the biota, is illustrated in Fig. 1.10.

A. Knowledge Base in Fishery Science

Fishery science is little more than a century old (Benson, 1970). The study of artificial habitats is a relatively young branch within it. A review of English language research literature on artificial reefs through 1983 identified 413 references (Bohnsack and Sutherland, 1985), of which over 75% were written after 1970. For the specific subject of fish-aggregating devices, 86% were written after 1969 according to Vega (1988) in a brief review of 211 articles. Both surveys identified published areas of emphasis as general program descriptions, structural design of habitat, biology of associated organisms, and monitoring and assessment (Table 1.2).

Bohnsack and Sutherland (1985) principally reviewed scientific papers, as opposed to numerous popular articles (see Chapter 2), and found that program histories for particular geographic areas (16%) and general descriptive articles about reefs (16%) were most common. They also determined that 31% of the articles were in peer-review journals, and 19% used experimental approaches of varying sophistication. These authors concluded that there was a "general lack of fundamental knowledge concerning optimum design criteria, location, and size of reefs," and made 29 recommendations for future studies.

Vega (1988) also noted the qualitative nature of many studies in a review that located primarily articles in English (87%) and French (11%). The earliest article he found on FADs was published in 1924, whereas the oldest publications identified by Bohnsack and Sutherland (1985) date from the

TABLE 1.2
Subjects Addressed by Publications on Artificial Habitats

Topic	Frequency of literature coverage	
	Artificial reefs	Fish aggregating devices
Design and Construction	19%	36%
Monitoring and Biology	39%	25%
General	34%	22%
Economics and Sociology	4%	6%
Other	4%	11%

[a]From Bohnsack and Sutherland, 1985; Vega, 1988.

1930s. These latter authors targeted marine situations, whereas Brown (1986) investigated the pioneering freshwater work of the ichthyological team of Hubbs and Hubbs in the early 1930s.

Subsequent to the literature cited in the two reviews above, additional technical articles have started to close the gap between the practice of habitat manipulation and its actual understanding and quantitative explanation. Notably, the *Bulletin of Marine Science* devoted two issues to artificial habitats in 1985 and 1989 (Buckley *et al.*, 1985; Seaman *et al.*, 1989). They contained 31 and 51 peer-reviewed articles, respectively, and a combined total of 45 abstracts. This rate of publishing compares favorably with the annual average of 22 papers on FADs determined by Vega (1988) for the 1980s, and the 33 per year on reefs in the early 1980s counted by Bohnsack and Sutherland (1985).

Review of selected Japanese literature has been provided by Mottet (1981) and Grove and Sonu (1983), among others, as a means of increasing access to a significant body of work. The former cited over 120 Japanese language articles, while the latter work drew upon seven principal sources of information. Translations of 20 Japanese-language articles in Vik (1982) focus primarily on reef design. As indicated in Chapter 2, the experience of Japan and the United States has formed the basis for much of the world's published technical literature, especially for benthic reefs. Inspection of literature that reports experience in other nations reveals a significant reliance on these references, particularly those in English.

B. Research Issues and Techniques

Creation of artificial habitats is a field that originated with fishermen. In its early phase as a scientific subject it by and large involved biologists, who often only could be concerned with construction but not ecological analysis. Hence, recent reviews of the biological literature have identified gaps and limitations in the research data base globally. Bohnsack and Sutherland (1985), for example, concluded that inadequate long-term monitoring of artificial reefs precluded explanation of their function biologically. Concern for insufficiency of biological data was expressed by Duval (1986), also. Others (Mottet, 1981; Grove and Sonu, 1983) concluded that evidence in the Japanese literature for increased fish production was not conclusive. Moreover, a lack of research on economic performance led Milon (1989) to conclude that "the practical significance of this neglect is that fishery resource managers have no objective basis to determine whether artificial habitat development has been an efficient use of society's resources and the consequence of these developments for fishery productivity" (p. 836). In effect, such authors are raising a larger issue of whether perceptions of increased efficiency or productivity are consistently supported by factual information. Alternate viewpoints on this subject and scientific aspects of the responsible application of habitat technologies were presented by Meier *et al.* (1989) at the Fourth International Conference on Artificial Habitats for Fisheries in Miami, Florida (1987).

Until recently, the ecological processes that determine the assemblage of species and composition of biomass on and around a given structure were of little concern to the varied groups interested in catching fish at artificial structures, or interested in the plan or construction of the structures. The practical consideration has been that, for whatever reason, these devices "work" by promoting a larger or more efficiently harvested yield. A general model of the factors influencing fish biomass on artificial reefs was presented by Bohnsack and Sutherland (1985) for (1) increases of biomass, possibly resulting from attraction (through recruitment and immigration) and growth, and (2) loss of biomass via mortality (e.g., fishing), emigration, respiration, and reproduction. In a review of the limited literature concerning the importance of habitat-limitation and behavioral preference as an explanation for fish abundance at artificial reefs, Bohnsack (1989) proposed that fish attraction and production are not mutually exclusive but that a gradient (or continuum) exists for their relative contribution depending on local conditions. Rountree (1989) reviewed hypotheses concerning fish association with FADs, which appear less complex ecologically than reefs, and indicated the mechanisms of protection from predators for the prey that seek shelter at FADs. Chapters 3 and 5 treat the ecological and fishery aspects of reefs and FADs in detail.

As a prelude to the extensive material presented in the following chapters concerning research on artificial habitats, we note two methodological aspects of their scientific study. First, a growing reliance has been placed on *in situ* observations of habitats by self-contained underwater breathing apparatus (SCUBA) divers. This is particularly true in environments that are more tropical and where visibility is relatively high (Fig. 1.11). Secondly, much of the experimental work on artificial habitats has been done on relatively small structures, which frequently are made of concrete blocks. The similarities of such modular habitats, independently designed in different locations, are depicted in Fig. 1.12. Below we note priorities for research on these and other size systems.

VI. Issues and Themes

The use and study of artificial habitats is a maturing area of fishery science. The concluding section of this chapter briefly reviews some important themes, issues, and priorities pertinent to their future development.

Figure 1.11 One of a pair of scuba divers record direct observations at an artificial habitat site (from Brock and Norris, 1989). Diving with a "buddy" is an essential aspect of established safety practices mandated by international organizations such as UNESCO (from Flemming and Max, 1988).

Specifically, we address (1) information needed to document habitat function and performance, (2) opportunities to revise or expand systems of planning and information, and (3) new challenges and practices of management and design. Highlights from the limited number of published reviews and assessments are incorporated; the brevity of our discussion is intended to provide guideposts for the remaining chapters of this volume. We also seek to alert the reader to significant discussion, debate, and controversy among knowledgeable workers in this field.

A. Habitat Function and Performance Evaluation

Is there a fundamental distinction between natural habitat and artificial habitat? At some point in considering how habitats such as benthic reefs on the sea floor or midwater fish attractors in a lake actually "work," the issue of whether man-made aquatic habitats are indeed "artificial" arises. For example, common materials such as concrete modules become permanent additions to the underwater landscape. Over time, do they mimic natural habitats of comparable dimensions? The issue of functional equivalency is being addressed by ecologists involved in a variety of habitat creation or

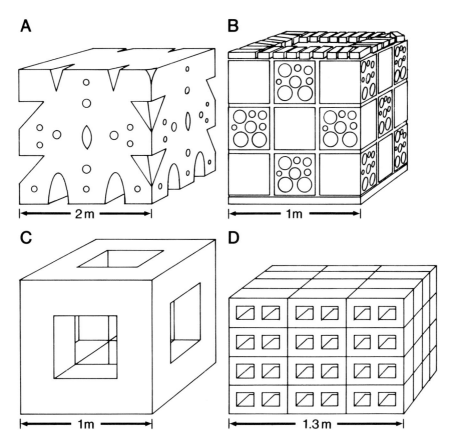

Figure 1.12 Modular concrete structures used in benthic reef experiments and research from different countries of the world reflect a convergence of design considerations. (Adapted from Bombace, 1989 **A**: Italy, 8 m³; Zahary and Hartman, 1985 **B**: USA, 1 m³; Chang, 1985 **C**: Taiwan, 1 m³; Hixon and Beets, 1989 **D**: Virgin Islands, ca. 1 m³).

restoration efforts, such as establishment of wetlands, and is appropriate to the consideration of the structures identified in this book in view of the financial investments and physical space they require in the environment.

In their review of the progress made in artificial reef research, Bohnsack and Sutherland (1985) recommended new priorities for studies that will further optimize reef design, size, and location as part of a broader effort to develop a "comprehensive theory on artificial reef operation and management" (p. 31). These authors summarized the literature that has reported rapid colonization of new reefs; biological functions of reefs that include shelter, feeding, spawning, and orientation for fishes; the tendency for fish abundance to be greater on artificial reefs than natural reefs; and the relatively greater importance of reef location, versus vertical relief, complexity, texture, and other influencing factors, which together add to the "success" of reefs. Nonetheless, these authors also stressed the need for new longer-term and larger-scale studies that include adequate experimental controls and replicates that will characterize primary and secondary productivity, energy flow, behavior, and other biological factors that as yet are unexplored or poorly understood. (See Chapters 3 and 6.)

Consistent with the conclusions of Bohnsack and Sutherland (1985), but based on an alternate approach, Steimle and Figley (1990) reported on a survey of managers and scientists in the United States. This sector also gave high priority to the need for new research. Ten broad areas (e.g., estuarine applications, community ecology) were described individually and collectively as the "key to the future of effective long-term fishery resource management" (p. 23). Steimle and Figley (1990) noted the importance of defining with greater precision the questions or problems to be answered, and suggested regional cooperation to address common research needs. The broad themes these authors identified for future research emphasis are the long-term overall (1) cost-effectiveness of habitat materials and designs, (2) benefits or effects of reefs on fish and shellfish populations and (3) fishery use of reefs.

Concerns about the performance and function of artificial habitats also extend to their economic impact. Milon (1989) expressed concern that conclusions in this area were premature due to a lack of well-designed and comprehensive assessments. He recommended predeployment studies to develop baseline data for use in evaluating attainment of the objectives of artificial habitats, and defined research themes for commercial and recreational fisheries that use artificial habitats. (See Chapter 7.)

Implicit in the preceding discussion is the need to develop research opportunities for defining the ecological structure and function of natural reef ecosystems, and also to characterize the failures of artificial habitats as a means of general understanding. The maturing of this overall subject area is clearly indicated by the formulation of a new generation of research issues and questions that requires a holistic and interdisciplinary approach.

B. Planning and Information Needs

Because much information about the deployment of artificial habitats is not reported in the technical literature, it has been suggested that there would be many advantages to establishing a worldwide information reporting system (Seaman et al., 1989; Steimle and Figley, 1990). In view of current technology, a computerized database for information storage and analysis would be highly desirable. Establishing such a database would require a great deal of cooperative effort and the examination of a number of factors to ensure maximum ease of use and compatibility of formats for the kinds of data gathered and the media used in recording. In assessing the feasibility of such an effort, questions specific to artificial habitats and data base format must be addressed. For example, could standard assessment and monitoring procedures be developed that would produce comparable data? Minimum information needs standard to all reef research projects were recommended by Bohnsack and Sutherland (1985), including reef size, composition, and age that should be recorded consistently. Further, computer operating system and data base application considerations pertinent to the largest number of users must be resolved. We believe this general issue should be taken up by fisheries interests in an international context. With such a system, the production of fishery species developed from artificial habitats located in different areas could be better assessed, and overall performance—using ecological, economic and engineering criteria—quantified. (See Chapters 2 and 5.)

Such a network would certainly enhance the exchange of information relative to overall planning of artificial habitats. With the establishment of a U.S. national reef plan (Stone, 1985), local and state reef plans were encouraged. Indeed, of 29 priorities recommended for the United States Atlantic coast, the five highest include the development of an "artificial reef policy and plan" by each state (Reeff et al., 1990). Such global planning would facilitate long-range investment decisions and perhaps allow thorough inventory of the actual contribution of artificial habitats to fisheries yields. Regional and international conferences are to be encouraged, and, as in recent years, continuing exchanges held in countries such as Sri Lanka, France, and the Soviet Union.

C. Enhanced Design and Management

The construction of artificial habitats may have a more profound effect on fishery management than might be envisioned from their relatively small contribution to global fish production. The slow and possibly slowing rate of growth of global fish harvest, rapid human population growth, and discord over allocation of fisheries resources, all portend a period of increasing uncertainty and conflict. The institutional arrangements coming into practice with regard to artificial habitat development may provide models for the broader institutional framework for fisheries allocation. Thus, *ad hoc* arrangements similar to those found in the Mediterranean, in which local communities deploy artificial habitat to impose a physical barrier to mechanized trawl gear and to provide a locally controlled and enforced area for regulated community fishing, may become more widespread. According to Bombace (1989) the opinions of Italian fishermen about artificial habitat deployment changed dramatically as their experience with artificial habitats proved profitable. Initially, their attitudes were indifferent or doubtful but became enthusiastic following increased catches and the exclusion of illegal trawlers from their local fishing grounds. Now, they actively promote the creation of more zones protected by artificial habitat structures.

In such a context where users of the resource system desire additional deployment of artificial habitat, scientists and managers must have an objective, quantitative basis for determining the biological carrying capacity of an area to support new habitats and fisheries. Both ecological and socioeconomic limits must be considered. Also, the materials and structures must be evaluated. Disposal of solid wastes, as a primary goal, has led to justifying the process as a means of secondarily enhancing fishing and fish populations. A progression of industrial and governmental concerns has sought to construct fish habitat from their waste or obsolete materials, particularly in the last 20 years. In response to solid waste disposal and landfill problems, automobile and truck tires and bodies were initially used. More recently, disposal of obsolete petroleum production platforms has been proposed and implemented on a limited basis. In the United States, for example, a governmentally sponsored "Rigs-to-Reefs" program was organized (Reggio, 1989). Also proposed is the utilization of solidified ash wastes from oil and coal-burning electrical-generating stations. Experimental reefs made from blocks of combustion waste mixed with concrete have been deployed in North America, western Europe, and Taiwan (e.g., Chang, 1985). Trace-metal leaching is one environmental aspect of oil ash reefs studied by Metz and Trefry (1988), who also identified the latest initiative for solid-waste disposal, namely municipal incinerator ash. Considerable care is essential to determine precisely the environmental and socio-economic costs, benefits, advantages, and constraints for each new material proposed as an artificial habitat.

Meanwhile, the challenge remains to integrate the physical design of structure with the biological requirements of organisms targeted by habitat deployment efforts. (See Chapter 4.) Bohnsack (1989) predicted what sort of fish species might be most amenable to biomass increases associated with creation of habitat. In turn, such species could be investigated to determine how additional structure might enhance survival, growth, or reproduction. The most extensive data base for matching habitat design with life history requirements comes from Japan. For example, a single layer of rocks on the sea bottom provides satisfactory substrate for seaweed production, but when piled to about one-half meter the layers of rock are used by sea urchins and abalone (Mottet, 1981).

In sum, as the technology of artificial habitats is applied over broader geographic areas, managers and practitioners alike will need increasingly sophisticated information that allows them to avoid past mistakes while seeking to achieve optimum results. Experimentation must expand to include full-scale artificial habitat systems, and international exchange of data and transfer of technology can be accelerated to use these systems in appropriate ways to enhance fishery and environmental management. Interests as diverse as the lay public and scientists will find in artificial habitats another window to the living underwater landscape that offers such a variety of benefits to society.

References

Agarwala, R. 1982. Planning in developing countries: Lessons of experience, Working Paper 576. World Bank, Washington, D.C.

Benson, N. G., editor. 1970. A century of fisheries in North America, Special Publication 7. American Fisheries Society, Washington, D.C.

Bohnsack, J. A. 1989. Are high densities of fishes at artificial reefs the result of habitat limitation or behavioral preference? Bulletin of Marine Science 44:631–645.

Bohnsack, J. A., and D. L. Sutherland. 1985. Artificial reef research: A review with recommendations for future priorities. Bulletin of Marine Science 37:11–39.

Bombace, G. 1989. Artificial reefs in the Mediterranean Sea. Bulletin of Marine Science 4:1023–1032.

Boy, R. L., and B. R. Smith. 1984. Design improvements to fish aggregating device (FAD) mooring systems in general use in Pacific island countries, Handbook 24. South Pacific Commission, Noumea, New Caledonia.

Brock, R. E., and J. E. Norris. 1989. An analysis of the efficacy of four artificial reef designs in tropical waters. Bulletin of Marine Science 44:934–941.

Brown, A. M. 1986. Modifying reservoir fish habitat with artificial structures. Pp. 98–102 in G. E. Hall and M. J. Van Den Avyle, editors. Reservoir fisheries management: Strategies for the 80's. Reservoir Committee, Southern Division, American Fisheries Society, Bethesda, Maryland.

Buckley, R. M., D. G. Itano, and T. W. Buckley. 1989. Fish aggregation device (FAD) enhancement of offshore fisheries in American Samoa. Bulletin of Marine Science 44:942–949.

Chang, K. 1985. Review of artificial reefs in Taiwan: Emphasizing site selection and effectiveness. Bulletin of Marine Science 37:143–150.

Ditton, R. B., and L. B. Burke. 1985. Artificial reef development for recreational fishing: A planning guide. Sport Fishing Institute, Artificial Reef Development Center, Washington, D.C.

Duval, C. 1986. Bilan sur les données biologiques de récifs artificiels imergés dans le monde. Pp. 15–20 in Journée d'études sur les aspects scientifiques concernant les récifs artificiels et la mariculture suspendue. Commission Internationale pour l'Exploration Scientifique de la Mer Méditerranée, Monaco.

Everhart, W. H., and W. D. Youngs. 1981. Principles of fishery science. Comstock Publishing Associates, Cornell University Press, Ithaca, New York.

Falundi, A. 1973. Planning theory. Pergamon Press, New York.

Flemming, N. C., and M. D. Max, 1988. Code of practice for scientific diving: Principles for the safe practice of scientific diving in different environments, UNESCO Technical Papers in Marine Science 53. Scientific Committee of the Confédération Mondiale des Activités Subaquatiques, Paris.

Friedman, J. 1987. Two centuries of planning theory: An overview. Princeton University Press, Princeton, New Jersey.

General Fisheries Council for the Mediterranean. 1986. Report of the Technical Consultation on Open Sea Shellfish Culture in Association with Artificial Reefs, FAO Fisheries Report 357. Food and Agricultural Organization of the United Nations, Rome.

Grove, R. S., and C. J. Sonu. 1983. Review of Japanese fishing reef technology, Technical Report 83-RD-137. Southern California Edison Company, Rosemead, California.

Hixon, M. A., and J. P. Beets. 1989. Shelter characteristics and Caribbean fish assemblages: Experiments with artificial reefs. Bulletin of Marine Science 44:666–680.

Hueckel, G. J., R. M. Buckley, and B. L. Benson. 1989. Mitigating rocky habitats using artificial reefs. Bulletin of Marine Science 44:913–922.

Japan Coastal Fisheries Promotion Association (JCFPA). 1986. Artificial reef fishing grounds construction planning guide. Technical report, Fisheries Development and Promotion Department of Japan Fisheries Agency, and National Research Institute for Fisheries Engineering, Tokyo.

Japan National Coastal Fishery Development Association. 1987. The ocean with unlimited possibilities: Promotion of the 3rd Coastal Fishing Ground Development Project, Tokyo.

Manar, L. 1966. Progress, 1964–1965, Circular 243. Biological Laboratory, U.S. Department of Interior, Honolulu, Hawaii.

McGurrin, J. M., R. B. Stone, and R. J. Sousa. 1989. Profiling United States artificial reef development. Bulletin of Marine Science 44:1004–1013.

Meier, M. H., R. Buckley, and J. J. Polovina. 1989. A debate on responsible artificial reef development. Bulletin of Marine Science 44:1051–1057.

Metz, S., and J. H. Trefry. 1988. Trace metal considerations in experimental oil ash reefs. Marine Pollution Bulletin 19:633–636.

Milon, J. W. 1989. Economic evaluation of artificial habitat for fisheries: Progress and challenges. Bulletin of Marine Science 44:831–843.

Mottet, M. G. 1981. Enhancement of the marine environment for fisheries and aquaculture in Japan, Technical Report 69. Washington Department of Fisheries, Olympia.

Murray, J. 1989. A policy and management assessment of Southeast and Mid-Atlantic artificial reef programs, UNC-SG-WP-89-3. UNC Sea Grant College Program, University of North Carolina, Raleigh.

Phillips, S. H. 1990. A guide to the construction of freshwater artificial reefs. Sport Fishing Institute, Washington, D.C.

Pramokchutima, S., and S. Vadhanakul. 1987. The use of artificial reefs as a tool for fisheries management in Thailand. Pp. 427–441 in Indo-Pacific Fishery Commission Symposium on the Exploitation and Management of Marine Fishery Resources in Southeast Asia, RAPA Report 1987/10. Regional Office for Asia and the Pacific, Food and Agriculture Organization of the United Nations, Bangkok.

Reeff, M. J., J. Murray, and J. McGurrin. 1990. Recommendations for Atlantic state artificial reef management, Report 6. Atlantic States Marine Fisheries Commission, Recreational Fisheries, Washington, D.C.

Reggio, V. C., Jr., compiler. 1989. Petroleum structures as artificial reefs: A compendium. Fourth International Conference on Artificial Habitats for Fisheries, Rigs-to-Reefs Special Session, November 4, 1987, OCS Study/MMS 89-0021. U.S. Minerals Management Service, New Orleans, Louisiana.

Rittel, H. W. L., and N. W. Webber. 1973. Dilemmas in a general theory of planning. Policy Science 4:155–169.

Rountree, R. A. 1989. Association of fishes with fish aggregation devices: Effects of structure size on fish abundance. Bulletin of Marine Science 44:960–972.

Seaman, W., Jr., R. M. Buckley, and J. J. Polovina. 1989. Advances in knowledge and priorities for research, technology and management related to artificial aquatic habitats. Bulletin of Marine Science 44:527–532.

Shomura, R., editor. 1977. Tuna bait-fishing papers, Circular 408. U.S. National Marine Fisheries Service, Honolulu, Hawaii.

Steimle, F., and W. Figley, compilers. 1990. A review of artificial reef research needs, Recreational Fisheries Report 7. Atlantic States Marine Fisheries Commission, Washington, D.C.

Stequert, B., and F. Marsac. 1989. Tropical tuna surface fisheries in the Indian Ocean: Floating objects and tuna purse seining, FAO Fisheries Technical Paper 282. United Nations Food and Agricultural Organization, Rome.

Stone, R. B., compiler. 1985. National artificial reef plan, NOAA Technical Memorandum NMFS OF-6. U.S. Department of Commerce, Washington, D.C.

U.S. National Research Council. 1988. Fisheries technologies for developing countries. Report of an ad hoc Panel of the Board on Science and Technology for International Development. National Academy Press, Washington, D.C.

Vega, M. J. M. 1988. Who's working on fish aggregating devices? Naga, The ICLARM Quarterly (October).

Vik, S. F., editor. 1982. Japanese artificial reef technology: Translations of selected recent Japanese literature and an evaluation of potential applications in the United States, Technical Report 604. Aquabio, Inc., Belleair Bluffs, Florida.

Welcomme, R. L. 1972. An evaluation of the acadja method of fishing as practised in the coastal lagoons of Dahomey (West Africa). Journal of Fish Biology 4:39–55.

Yoakum, J., W. P. Dasmann, H. R. Sanderson, C. M. Nixon, and H. S. Crawford. 1980. Habitat improvement techniques. Pp. 329–403 in S. D. Schemitz, editor. Wildlife management techniques manual. The Wildlife Society, Washington, D.C.

Zahary, R. G., and M. J. Hartman. 1985. Artificial marine reefs off Catalina Island: Recruitment, habitat specificity and population dynamics. Bulletin of Marine Science 37:387–395.

2

Artificial Habitats of the World: Synopsis and Major Trends

R. B. STONE
Recreational and Interjurisdictional
Fisheries (F/CM3)
National Marine Fisheries Service
Silver Spring, Maryland

J. M. Mc GURRIN
Atlantic States Marine
Fisheries Commission
Washington, D.C.

L. M. SPRAGUE[1]
Agriculture and Rural
Development Department
World Bank
Washington, D.C., and
Department of Fisheries and Aquaculture
University of Florida
Gainesville, Florida

W. SEAMAN, JR.
Florida Sea Grant College Program, and
Department of Fisheries and Aquaculture
University of Florida
Gainesville, Florida

I. Introduction

Imagine the excitement perhaps thousands, and certainly hundreds of years ago, as fishermen discovered that a floating object or an underwater structure enhanced their catches. Since then, efforts to influence the behavior and abundance of aquatic organisms have flourished. Although many enhancement practices arose independently and in isolation, they evolved common elements of design. As global interest in these techniques continues to

1. Retired from World Bank.

grow, it is useful to recount some of the historical patterns of development in the field.

Most efforts to deploy artificial habitats are relatively recent; only a few nations have been involved for more than two decades, and technical documentation in the literature has generally lagged. In a few areas though, certain basic methods have been followed for centuries. This chapter summarizes, with some limitations, the global pattern of development for this subject.

To complement cursory published information for many regions, we also present case histories of the two better documented, more extensive, and nationwide approaches to artificial habitat development, namely for Japan and the United States. The highlights of some reports are presented to avoid repetition of well-documented material. However, we also analyze certain events to identify trends that have not been discussed previously in a single publication.

Finally, by indicating the philosophy of some systems of artificial habitat development and ownership, this chapter offers comparative information that is intended to be useful in understanding their evolution. The concerns of managers and scientists are integrated with those of the artisanal, commercial, and recreational fishing interests whose experiences form some of the foundation for this field.

II. Scope and Extent of Global Artificial Habitat Development

There is clearly a broad spectrum of people interested in using the artificial habitats that are the focus of this book, ranging from private actions of individuals to national government sponsorship of large programs. This section reviews the more formally organized activities, based principally on documentation in the published technical literature. Because of the growing acceptance and application of these technologies, the list of involved countries, provinces, and local governments, plus private interests, continues to expand. Comprehensive documentation of program organization and results is limited, however.

Despite the worldwide popularity of this endeavor, no global computerized data base is available. While this may not be surprising, there is also typically a lag between program implementation and actual preparation of published reports, except perhaps for administrative or nontechnical documents that do not receive wide circulation. Therefore, a comprehensive review of the scope and extent of this field is compromised. (A similar situation exists for research literature in this field, although a significant number of reviewed articles have appeared since 1985. This is addressed in Chapter 1, which also presents illustrations and an overview complementary to the following narrative.)

A. Synopsis

Approaches to organizing efforts to deploy artificial habitats initially may appear to be as diverse as the materials and structures used. However, by analyzing the interests responsible for promoting, organizing, or using these practices, it is possible to discern different categories of formal and informal efforts.

Variations in geographic extent and involvement of different sectors are discussed. Activities are reviewed on the basis of the participation of national governments, their geographic extent and objectives, and the roles of management and user group sectors. (A complementary approach is taken in Chapter 5, which discusses representative data on fishery and biological impacts and performance.)

This synopsis is principally a review of coastal and oceanic man-made habitats, due to the paucity of literature for freshwater situations. Also existing reports of long-established practices of localized effect need not be repeated here. (See, e.g., Phillips, 1990, who reviews construction techniques, and D'Itri, 1985, who presents the proceedings of a conference that include the Great Lakes of North America.) Much of the recent technological and scientific focus in the field of artificial habitats has been in the marine sector. As evident from the material in this book, a number of techniques can be applied in both fresh and salt water environments (Fig. 2.1).

1. National Programs

The earliest major national governmental intervention in this field began about 60 years ago in the coastal waters of Japan (detailed information is provided in the case study later in this section). An instructive commentary on the relatively short history of this global field is that the pioneering Japanese workers in this area are active professionally or just now retiring.

Subsequently, national plans and programs—or major regional efforts within a country—have developed in the last two decades, or are pending, in the United States, Thailand, India, Taiwan, Malaysia, Australia, and the South Pacific Islands.

The level of governmental commitment to so-called "national" programs can vary markedly. The largest financial obligation of a federal government is in Japan, with funding in recent years of billions of yen annually (Yamane, 1989). Here, significant government support for construction costs has led to establishment of an industrial infrastructure, while a large research program

Figure 2.1 The log crib is among the structures that can be used in freshwater and coastal marine settings. In more temperate areas, this structure can be constructed in winter on the frozen surface of a lake, to sink to the bottom when the ice thaws in spring.

also has evolved. Geographically, nearly 10% of Japan's ocean shelf has received what Yamane refers to as "improvements." No other federal government is as heavily involved as Japan (see case study, section IV).

Elsewhere, the variety of artificial habitat building efforts reveals that most national government involvement is limited financially or in geographic scope. Thus, definition of what constitutes a national program is somewhat arbitrary. Here, we include countries whose principal fishery agency or overall natural resources ministry allocates some minimum budget to artificial habitats, or has a planning document to guide expenditures and construction.

National plans and national programs are not synonymous. That is, a national government (1) may have a plan and fund deployment (e.g., Japan); or (2) it may have a plan but no, or only minimal, funding (e.g., U.S.A); or else (3) it has no plan (yet) but is committing modest funding to experimental or full-scale deployment of habitat (e.g., India). Such national efforts may not exclude other (nonfederal) organizations from participating. Such collaboration—between different levels of government or among different private interests—can be very productive.

Evolution of habitat programs at a national level follows a variety of courses. Whereas early central leadership was exerted in Japan, in other nations either the development of a plan or the distinctly separate act of providing funds for activities happened in response to the initiative of other interests. The sources of funding or availability of other resources to support habitat programs are variable. In India, for example, two institutional re-search centers demonstrated benthic and surface habitat structures on the Madras coast beginning in 1986 (Sanjeeva Raj, 1989). After workshops popu-larized the practices in coastal villages, the Department of Ocean Develop-ment and the Department of Biotechnology of the Indian government asked for proposals to develop a large-scale project. The state governments of Tamil Nadu and Pondicherry also have expressed interest. In the United States, also, state governments have been a primary factor in deploying ar-tificial structures in marine and freshwater environments, as discussed later in detail. Their collective efforts are far greater than the minimal effort of the U.S. federal government.

The motivation for a national program can vary. Since 1978, the Depart-ment of Fisheries of Thailand has constructed 34 artificial concrete and tire reefs as part of a marine conservation program and also to enhance coastal fishing while reducing conflict between artisanal and commercial fishermen (Pramokchutima and Vadhanakul, 1987). Improved recreational fishing, meanwhile, is the principal objective of Australian and U.S. artificial habi-tats. In Taiwan, a three-fold purpose of fishery enhancement, environmental mitigation, and disposal of combustion ash has been proposed as part of a master plan for artificial reefs to be built in conjunction with electric power plant construction in the northern region of the country (Chang and Shao, 1988; Shao, 1988). Assessment of habitats in Malaysia is just beginning, so documentation does not yet exist for the tires, derelict vessels, and other structures deployed by the Department of Fisheries.

According to Buckley et al. (1989), by the early 1980s virtually all coun-tries and territories in the South Pacific Ocean had instituted or planned programs to use fish-aggregation devices (FADs) to enhance pelagic fisher-ies. For example, American Samoa began a program in 1979, which has pro-gressed through design improvements in FAD systems. Gates (1990) re-ported that 431 FADs were deployed in 15 South Pacific countries in the period between 1984–1990, compared to 600 deployed before 1984. Favor-able catch and effort data for FADs were recorded by Buckley et al. (1989). The genesis of this technology in Hawaii and the Philippines is described by Preston (1990).

Artificial habitats are also created principally through the efforts of local private interests, as a precursor to government involvement. They may be-gin completely independent of federal, provincial, or local government ac-tivity or interest. In the Maldives, for example, FADs were introduced in 1981 and their success has prompted suggestion of a more formal program (Naeem, 1989). International development agencies also participate, such as the World Bank, the Food and Agricultural Organization of the United Nations, and Bread for the World.

Governmental and private initiatives are not necessarily mutually exclusive. For example, in Australia both private recreational fishermen and agency scientists are involved in deployment of habitats. Since 1965, all areas of Australia have witnessed habitat deployment (Pollard and Matthews, 1985). Numerous private U.S. sportfishing clubs may receive permits to build reefs in individual state coastal waters, sometimes in collaboration with local or state governmental agencies. Interests in both nations continue to rely primarily on materials of opportunity. (See Fig. 1.1 and other figures in this volume for drawings and pictures of common artificial habitats).

2. Regional Approach

In many other nations, efforts have been made with a more limited geographic range or on a feasibility basis. Thus, we do not consider these activities "national" in scope, although they may be the initial step in the evolution of a country's official program. In the following description, funding may or may not be federally derived, and the initiative may come from a variety of public and private interests.

Italy may have the oldest and most geographically widespread regional program, having started experiments with artificial habitats in the 1970s (Ardizzone *et al.*, 1989). In the Ligurian Sea, for example, the goal of restoration of the inshore ecosystems was implemented using artificial reefs to deter illegal trawling (Fig. 2.2), repair deteriorated coastline, and ameliorate sewage pollution (Relini and Relini, 1989). Of 34 artificial reefs planned in Italian coastal waters, 7 were completed and 15 were in progress as of 1987 (Bombace, 1989). Production of bivalves, including mussels (*Mytilus galloprovincialis*) and oysters (*Ostrea edulis* and *Crassostrea gigas*), from habitats de-

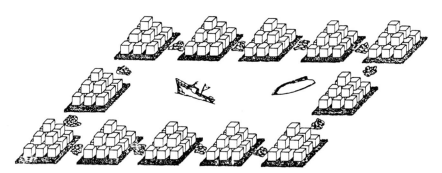

Figure 2.2 This "reef zone" in the Adriatic Sea of Italy is representative of ways that artificial habitats can serve multiple purposes of fishery management (from Bombace, 1989).

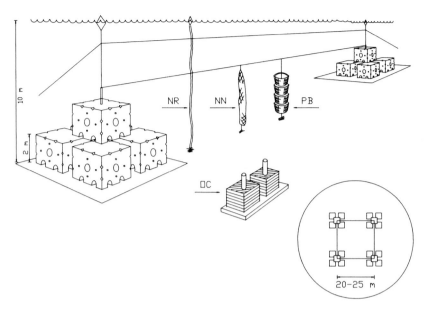

Figure 2.3 Two concrete modules for fish and shellfish habitat, with nylon rope (NR), nylon net (NN), oyster collectors (OC), and plastic baskets (PB) used for mussel and oyster culture in coastal areas of the Adriatic Sea in Italy (from Fabi *et al.*, 1989). (Insert shows the spacing of clusters of concrete block).

signed for a variety of species and purposes (Fig. 2.3) prompted fishermen to apply for continued development of new habitats (Fabi *et al.*, 1989).

Elsewhere in the Mediterranean basin (Bombace, 1989), contemporary efforts have taken place at several sites in France, beginning with deployment of concrete blocks and tires in 1970, and then concrete modules and construction materials in the 1970s and 1980s. In Monaco, emphasis is on the Monaco Underwater Reserve created in 1977 as part of an environmental restoration project to rehabilitate flora and fauna (De Bernardi, 1989). Massive reinforced concrete-block structures have been colonized by crustaceans (*Palinurus vulgaris*) and mussels, and finfish reproduction (by the scorpion fishes, *Scorpaena porcus* and *S. scrofa*) has been observed.

Also, according to Bombace (1989), an experimental reef is planned for a marine reserve in Spain. Of special prominence in France, Italy, Spain, and Bulgaria are surface and subsurface floating structures used for mussel and oyster culture. Based on reports of harvests from these areas, this practice is spreading to Greece and Algeria (Bombace, 1989). Use of structures in areas of high primary productivity and in habitat restoration in the Mediterranean is addressed by Bellan-Santini (1982).

In terms of significance of landings, FADs used regionally in the Philippines support one of the largest fisheries in the world based on artificial habitats. Tuna catches under 10,000 metric tons in 1971 increased to 125,000 metric tons in 1976, with peak production of 266,000 metric tons in 1986; these increases are attributed to FAD usage (Aprieto, 1988). More information is presented in Chapter 5 of this volume.

3. Local and Independent Efforts

Some of the oldest applications of artificial habitats have been practiced by individuals or local communities, without benefit of formal development programs or subsidies from organizations concerned with natural resources or fisheries management. In the central Mediterranean, for example, FADs known as *kannizzati* (Malta) and *incannizati* (Sicily) have been used for centuries. Deployment of long lines of connected cork blocks (shorter than one meter) is described by Bombace (1989). Historical usage of FADs in the Philippines was reported by the U.S. National Research Council (1988).

In western Africa, small bundles of branches and brush are deployed in various configurations collectively known as *acadja*. They have been used for over a century, with present practices in West Africa (Dahomey), Cameroon, and Nigeria described by Welcomme (1972); and he indicated that acadja are employed in coastal lagoons and lakes. Among the freshwater, brackish and marine fishes captured, tilapia (family Cichlidae) predominates in the Republic of Dahomey (Welcomme, 1972). Brush parks have been used elsewhere in Africa, including the Ivory Coast, Ghana, Togo, Sierra Leone, and Madagascar, as well as in Sri Lanka, Bangladesh, China, and Ecuador (Kapetsky, 1981).

A number of countries have pursued low-cost artisanal fishing strategies to serve one or more complementary purposes, including the following: restoration or replacement of degraded coral reefs, seagrass beds, and mangrove forests; creating new fisheries; reducing overcapitalization and overexploitation; and reducing fishing gear conflicts in existing fisheries. The Philippines, Jamaica, Cuba, Mexico, India, Indonesia, Sri Lanka, Maldive Islands, and Papua New Guinea are among the nations that have deployed floating and benthic structures that include bamboo rafts, brushpiles, coconut fronds, steel drums, and discarded tires.

Recent initial deployments of artificial habitat on a pilot basis, or experimentally, reflect widespread dissemination of technical information in this field, and the generally favorable accounts of experiences from areas with established efforts. On the Pacific coast of Central America, the first artificial reef in Costa Rica was established in 1984 in an embayment of the Gulf of Nicoya, and by 1986, 5000 car and truck tires had been deployed (Campos and Gamboa, 1989). On the Atlantic coast of Guatemala, an estuarine reef was built in 1982 (Bortone et al., 1988).

The existence of 150 North Sea petroleum production platforms on the continental shelf of the United Kingdom prompted an evaluation of their potential role in fisheries by Picken and McIntyre (1989). The relatively large distances offshore of most structures place them beyond fishing access, but their potential role as relocated inshore habitats is recognized.

Early in the 1980s, four small modules of tires bound by steel and fiberglass bars were deployed off Haifa, Israel to determine the feasibility of enhancing fisheries in the area of the Mediterranean Sea that is lowest in primary productivity (Spanier et al., 1985). In Kuwait three tire modules were deployed in 1981, the first artificial habitats reported in the Arabian Gulf (Downing et al., 1985).

From their ancient use in limited areas of the world, artificial aquatic habitats have proliferated globally in marine waters. Their usage in freshwater is less widespread. In both environments, earlier efforts are better documented in the literature. However, the literature typically emphasizes descriptions of physical aspects and deployment of habitats, and often presents only minimal information about species and harvest levels.

B. Trends Based on Limited Data Sets

This section supplements the preceding information, which reflects the contents of a limited number of technical publications regarding efforts in different countries and regions. Here, we use two surrogate measures to quantify trends in habitat development: (1) the records of international artificial habitat conferences, and (2) citations from bibliographies of artificial reef literature. Briefly, at the most recent international symposium held in Miami, Florida (1987) reports from 19 nations were delivered, while over 35 nations are reported in literature compiled in a scientific bibliography.

1. International Conferences

Changes in the levels of interest and activity in artificial habitats are reflected by topics and participation at four conferences that are billed as international in scope. They were held in Houston, Texas in 1974 (Colunga and Stone, 1974), 1977 in Australia, 1983 in Newport Beach, California (Buckley et al., 1985), and again in the United States in Miami, Florida,

1987 (Seaman *et al.*, 1989). Published proceedings from the first conference emphasized descriptive studies and reports of programmatic efforts, with emphasis on U.S. experiences at particular sites. However, information from the two most recent international conferences has been published as peer-review articles by the *Bulletin of Marine Science* (Seaman *et al.*, 1989; Buckley *et al.*, 1985), as a reflection of a growing body of increasingly sophisticated research. Emphasis has shifted from anecdotal and technical accounts of habitat construction at local sites to more quantitative analysis of biological and engineering performance.

The increasingly global character of the programs, now formally designated as the "International Conference on Artificial Habitats for Fisheries," is reflected in Table 2.1. Dramatic increases in attendance, presentations, and representation from nations other than the host country were observed for the 1987 conference. A more detailed analysis of the four programs is made by Seaman (in press). Although international participation in these conferences may be limited by cultural, financial, and linguistic barriers, we believe that the increase from 4 countries participating in 1974 to 26 coun-

TABLE 2.1
Program Statistics for International Conferences Dealing with Artificial Aquatic Habitats

	Year			
	1974[a]	1977[b]	1983[c]	1987[d]
Number of Sponsors	3	1	12	22
Number of Nations Represented by Attendees	4	17	7	26
Number of Nations Represented in Presentations	4	4	7	19
Number of Papers (Oral)	41	17	43	94
Number of Posters	—	—	14	37
Attendance	Ca. 250	?	150	349
Publication of Proceedings (No. Articles + Abstracts)	Gray Literature (32 + 0)	None	Journal (30 + 26)	Journal (51 + 19)

[a] International Conference on Artificial Reefs (USA).
[b] Artificial Reefs Symposium, 5th World Underwater Congress (Australia).
[c] Third International Artificial Reef Conference (USA).
[d] Fourth International Conference on Artificial Habitats for Fisheries (USA).

tries by 1987 reflects a growing worldwide interest in artificial habitats. (Greater participation at the 1991 conference is anticipated.)

2. Artificial Reef Bibliographies

Intensity of interest and activity in artificial habitats are measured here using bibliographies of reef citations, despite certain limitations. Two computerized bibliographies compiled by workers in the United States were analyzed. There is, of course, a bias favoring papers translated into English, papers that cover marine topics, or articles that met other specific requirements defined for each bibliography.

The first bibliography contains over 2500 references (Stanton et al., 1985). Diverse but related topics were identified, such as biofouling, effects of oil platforms on the environment, and breakwater engineering. The literature search was run on 24 reference data bases through the DIALOG® Information Retrieval Service. Even though the authors deleted historic and popular literature (e.g., news clippings), the number of citations increased five-fold over the number in a 1973 artificial reef bibliography by Steimle and Stone.

The second bibliography (Reeff and McGurrin, 1986) was produced by the Sport Fishing Institute (SFI). It was compiled in conjunction with the previous bibliography for use by the SFI Artificial Reef Development Center. It overlaps somewhat, but does not include the citations that are indirectly related to habitat development (e.g., breakwater construction, *de facto* reefs such as active oil platforms, bridges, and tunnels). It focuses only on constructed habitat, and its nearly 2300 references include popular articles and other gray literature.

For purposes of the present analysis, both bibliographies were edited and combined into a single data base of 2291 references. Numerical analysis of the marine and freshwater citations by country indicates worldwide interest in artificial habitat development (Table 2.2). Of the references analyzed, 1097 cited specific habitat development in 39 countries through 1986.

Clearly the data in Table 2.2 are to be considered relative. The 91 Japanese citations, for example, are a low estimate because little of the information about their reef building efforts is translated into English and would not be cited in these bibliographies. Similar underestimates can be found for Canada, which has an extensive number of freshwater stream improvement projects. Thus, the number of citations for each country does not represent a comprehensive or absolute index of artificial habitat development, but rather should be used as a relative gauge of activity. (See Table 1.1 for comparative information.)

To address the need for more information on freshwater environments

TABLE 2.2

Number of Country Citations from Two Combined Bibliographies of Artificial Habitat Literature

	Subject	
Region/Country	Saltwater	Freshwater
North America		
USA	743	112
Canada	3	11
Mexico	2	—
Central & South America		
Venezuela	2	—
Costa Rica	1	—
Nicaragua	—	1
Atlantic/Caribbean		
Virgin Islands	11	—
Puerto Rico	7	—
Bahamas	5	—
Bermuda	3	—
Cuba	2	—
Jamaica	2	—
Barbados	1	—
Africa/Indian Ocean		
South Africa	2	—
West Africa	1	—
India	1	—
Kuwait	1	—
Asia		
Japan	91	2
Philippines	6	1
Taiwan	5	—
Guam	3	—
Korea	2	—
Thailand	1	—
Indonesia	1	1
China (Hong Kong)	—	1
South Pacific		
Australia	14	2
New Zealand	2	1
Pago Pago/American Samoa	2	—
Europe/Mediterranean		
Italy	12	—
France	9	2
Germany	3	—
Israel	3	—
United Kingdom	2	1
Belgium	2	—
Soviet Union	2	11
Netherlands	1	—
Poland	1	—
Monaco	1	—

including stream improvement, a separate reference (Duff and Gnehm, 1986) was consulted. It contains 1106 entries of published and unpublished references related to resident and anadromous salmonids, and coldwater and warmwater fish species. Citations from 1933 to 1985 cover a geographic area from North America, Europe, and New Zealand. Because the bibliography is not annotated, it was not possible to systematically identify activities by country. Hence, this data base could not be combined with the other two bibliographies for analysis.

III. Philosophical Bases for Artificial Habitat Activities

The basic philosophical ideas about property rights in the coastal ocean appear to have determined the manner in which societies have shaped fisheries policy and conducted the development of artificial habitats in the coastal zone. It appears that three ideas about property rights, or variations of them, have shaped the development of present worldwide artificial habitat deployment practice. By design we do not present them in the formal and legalistic way typical of many discussions about law of the sea issues. (The reader may wish to consult Eckert [1979] for such material.)

The first is the idea that individuals, or groups of individuals, either own title to specific coastal areas or have certain tenurial rights to use specific areas of coastal ocean, or specific rights to specific fisheries. Such rights convey the right to determine what is done in the area, who does it, and when. For example, full tenure[2] to the resources and their use in certain coastal areas is certainly largely responsible for the dynamic development and success of artificial habitat deployment in Japan.

The second basis, a variation of the first, is characterized as *de facto* tenure. This refers to practical rights to the use of limited coastal areas, without the force of law, but with customary use, and local enforcement. In the Philippines and Indonesia, some island states and some other countries, a system of *de facto* tenure is in effect in many coastal areas. In such situations, local village councils or their equivalent set forth use rights, management practices, and enforcement measures appropriate to the area and circumstances.

The third concept is the idea that the oceans, seas, and bays—from an area of tidal influence seaward, up to some arbitrary limit, often 12 nautical miles, and extended for economic activities up to 200 nautical miles—are

---

2. Tenure of this type does not convey "title" to the physical area but serves the same purposes from a policy point of view.

the common property of the coastal state and its citizens. For example, in the United States, control and use of the coastal area is the responsibility of the state on behalf of the populace.

IV. Case Studies:
Japan and the United States

Even though detailed histories for most countries are unavailable, we used existing literature about the evolution of artificial habitat activities in the two most active nations, the United States and Japan, to highlight similarities and contrasts that may reflect extremes in the general evolution of habitat programs on a worldwide basis. Whereas the early histories of habitat building in Japan and the United States show certain similarities, present efforts concerning the use of materials and degree of planning exhibit sharply contrasting approaches. Extensive freshwater literature for the United States exists but not for Japan.

The Japanese government is actively involved in reef construction activities through its fishery agency subsidy program. The agency is involved in planning and guidance, and provides substantial funding for those projects that use government-certified reef products. Underpinning the Japanese system is a political approach that provides use rights to those who construct and deploy reefs. These rights convey the sole control of the harvest and use of the fishery resources around their structures (C. Sonu, personal communication).

In the United States most marine and freshwater habitat construction activities are carried out by state and local governments, with only general guidance and minimal funding provided by the federal government. Here reef development is incorporated into a common-property allocation system.

A. The Japanese Experience

The artificial habitat concept was used earlier and in other places, but the Japanese appear to be the first nation to systematically record their activities beginning about the late 18th century. This account addresses coastal and oceanic habitats.

1. History

Three phases of development may be identified in the history of the Japanese experience. The first was characterized by the small-scale application of artisanal reefs, depending mainly on materials of opportunity as reef components and transfer of technology by word of mouth. This phase dates back to at least the Kansei era (1789–1801) and probably earlier (Ino, 1974). Ino noted that in 1795, written reports documented that a fisherman of the Manzai village in the province of Awaji (Awaji Island, south of Kobe) fished by chance near a sunken ship and caught several thousand yellow spotted grunt, *Plectorhynchus cinctus*.

When the sunken ship deteriorated in 7 or 8 years and the fish disappeared, fishermen sank large wooden frames weighted with sandbags in waters approximately 40 m deep. Three months later, the fishermen netted a far greater number of fish near the new artificial reef than they used to catch around the sunken ship. In the decade that followed, they sank several hundred more units. Similar efforts occurred in other areas.

Ino (1974) also related the following story, taken from a stone monument in the village of Uoshima. During the Kansei era, a large junk loaded with 4962 bushels of rice struck some rocks and sank. The following year during the spring fishing season, village fishermen caught an extraordinarily large amount of sea breams (family Sparidae). Several decades later a village chief from the story inferred the connection between the sunken junk and the increase in the fishermen's catch and in 1876 sank straw bags filled with parched rice-bran and clay among existing rocks, hoping that sea bream fishing would improve. The effort was successful. Ino (1974) documented numerous other efforts during the early 1900s to enhance fisheries for a variety of species using many different materials.

The second stage began soon after World War II. To increase protein intake in the nation's diet, in 1954, the government actively promoted development of a distant water, high-seas fishing fleet and legislated artificial reef construction as a continuing national program backed by generous subsidies. During this time, the artificial reef program in Japan was dedicated to enhancing fish catches within or near existing natural reefs. This phase continued until the mid-1970s.

The typical traditional process of constructing artificial reefs in Japan was the so-called *Tsukiiso* (meaning man-made rocky shelf at the seashore). It consisted simply of dumping rocks and boats on the sea floor to enhance fish stocks at an existing natural reef. In 1954, under the government-subsidized program, the official name, *Gyosho* (meaning fishing reef), was established for reefs that are constructed with concrete blocks, as distinguished from the traditional *Tsukiiso*. Most of the materials used under this program were relatively small, consisting mainly of hollow cubes with typical dimensions ranging from 1.0 m to 1.5 m, and short pipes with typical diameters and lengths ranging between 1.0 and 1.8 m. The size of reefs constructed with these materials was also relatively small, averaging about 500 m³. (See illustrations in Chapter 4.)

In 1958, the program was expanded to allow construction of much larger reefs. *Oh-gata* (meaning large-scale reef) was introduced as a new category of artificial reef, whose size exceeds 2500 m³, in contrast to smaller reefs called *Nami-gata* (ordinary-scale). The large-scale reefs used the same types of concrete blocks as the ordinary-scale reefs, but they also included larger fabricated structures.

Differences between the large-scale and ordinary-scale reefs existed not just in size but also in intended use. The ordinary-scale reef was typically placed in a shallow area where it served only a single nearby fishing community. The large-scale reef was placed where it could serve more than one fishing community with overlapping fishing rights, and it was deployed further offshore and in deeper water than the ordinary-scale reefs. In spite of these differences, both the *Oh-gata* and *Nami-gata* had one important characteristic in common: their chief function was to supplement the existing natural reefs and fishing grounds, and they were usually much smaller in size than the natural reefs (C. Sonu, personal communication; also see Grove and Sonu, 1983, for detailed discussion).

By 1966, 721,065 ordinary artificial reef units (each equal to 1 m³ blocks), and 328,217 larger units (each equal to 1.4 m³ blocks) were installed. Between 1962 and 1970 the equivalent of 920,000 m³ of ordinary reef units were installed in 3427 localities, and 1,320,000 m³ of large-scale units were placed in 439 localities (Ino, 1974). The national expenditure provided by the government for habitat improvement during the period between 1952 to 1970 was approximately 10 billion yen.

The third stage of Japan's program began in 1974 with passage of the Coastal Fishing Ground Development Act (Yamane, 1989). In 1975 the Artificial Reef Fishing Ground Construction Program was initiated to use artificial reef technology to create new fishing grounds where there had been none. This concept of "creating" fishing grounds has evolved rapidly since about 1980, when the goal to convert the whole Japanese fishery from earlier conventional emphasis on distant-water fishing to a new emphasis on farming and resource management in home waters became explicit national policy.

Unlike the ordinary-scale or large-scale reef projects whose function it was to take advantage of existing natural reefs, the new program sought to create new fishing grounds. The artificial-reef fishing grounds were to be large, comparable in size to natural reefs, and usually at least 20 times larger than the earlier large-scale reefs. They were to be used to develop fishing grounds in relatively deep water, requiring the use of large fabricated structures for reef units. Total construction costs, for each fishing ground, were estimated to cost about U.S. $3 to 4 million. Government subsidies for the construction of artificial fishing grounds were 70% of the project cost, 60%

for the large-scale reefs, and 50% for the ordinary-scale reefs (C. Sonu, personal communication).

2. Government Involvement

Japan continues to pursue a policy of enhancing the productivity of coastal fisheries by increased investment in technology, innovation, and construction (Grove *et al.*, 1989). Administering a massive subsidy program, the Fishery Agency leads this trend. From 1976 to 1987, the government subsidy program provided funds to construct about 1.4 million m^3 of habitat structure. In the 1988 fiscal year budget, Japan proposed to spend close to U.S. $150 million to construct 2.2 million m^3 of fishing reefs.

Japan leads the world investment in fishing reef technology and marine ranching. For two six-year fishery plans (1976–1988) the national government committed U.S. $250 million and U.S. $500 million, respectively, for artificial reefs alone (Grove *et al.*, 1989). Subsequently, and in addition to developments reported by Mottet (1981), Grove and Sonu (1983), and Grove *et al.* (1989), oral information presented at the Fourth International Conference on Artificial Habitats for Fisheries indicated that heavy industry continues to be involved in fabrication of vast multistory structures, principally of concrete and steel. (See Fig. 1.5 in Chapter 1 and figures in Chapter 4.)

The Japanese Fishery Agency's authority is so pervasive that it influences all aspects of artificial reef planning, construction, and technology development. Government subsidies are limited to those projects (studies and construction) that use government-certified reef products. Since reef construction without government subsidy is economically impractical, subsidies allow the government to effectively control the quality of reefs including building materials, design, location, and construction. Analysis by Mottet (1981) indicated that the average cost to the government for a regular size reef (i.e., volume less than 2500 m^3) was U.S. $22,727, and a large reef (i.e., up to 50,000 m^3) was U.S. $327,000, with an additional 50% and 40% of costs provided by local organizations.

Recently, the government's efforts to refine the planning and design of artificial fishing reefs in Japan reached a milestone with the publication of two guides (Japan Coastal Fisheries Promotion Association, 1984, 1986). They specify revised standards that reef design and construction plans must meet to qualify for government certification. They are included in extensive discussion of Japanese engineering practices in Chapter 4.

A new multiyear plan, the Third Coastal Fishing Ground Development Project, proposes a 150% increase in the effort devoted to reefs in the current program. Whereas engineering and construction practices appear well-documented, aspects of ecology continue to need attention. A new ecological

Figure 2.4 This solar power station, part of the marine ranching program in Saiki Bay, Japan, exemplifies the advances emerging in artificial habitats technology. It provides electricity to underwater monitors and sensors (from Grove *et al.*, 1989).

emphasis seems to have been placed on structural modification to enhance nutrient upwelling. Evolution toward sophisticated "marine ranching" in coastal ecosystems is occurring (Figs. 2.4 and 4.16, see also Chapter 4.)

The comprehensive review by Mottet (1981) of recent practices in Japan presented information from over 120 Japanese language technical articles. Highlights of this analysis (Table 2.3) reflect a diverse set of practices to enhance harvest of coastal plant, invertebrate, and finfish species. In contrast to the extensive descriptions of physical aspects of Japanese artificial habitats (Fig. 2.5), Mottet (1981) concluded that there are "insufficient biological and economic data available for making informed judgements on the effectiveness and practicality of many of the operations." Mottet recognized the potential for simply shifting income from one locality to another if mi-

TABLE 2.3
Principal Structures Used to Enhance Fishery Species in Japan[a]

Material or Structure	Principal Applications
Rocks (in layers, piles, or in cages)	Substrate for seaweed
	Habitat for immature sea urchins and abalone
Substrate Blocks (concrete)	Seaweed holdfast
	Shelter for larval fishes and invertebrates
Breakwater Blocks (concrete)	Seaweed holdfast
	Stabilize seagrass and clam culture grounds
Chamber Structures (concrete cubes and cylinders)	Fish, abalone, and sea urchin production
Larger Chamber Structures (concrete, plastic, fiberglass, and steel frameworks)	Migrating fish attraction
Longline	Kelp growth and reproduction
Plastic Seaweeds	Shelter for small invertebrates and fishes
Bamboo Rafts	Pelagic fish attraction
Floating Devices	Pelagic and midwater fish attraction

[a] Source of data: Mottet, 1981.

grating fish are attracted to new artificial habitats, but she also documented productivity for newly established kelp forests and associated abalone and sea urchins.

B. The United States Experience

Artificial habitats have been used for over 100 years in the United States but have only recently been recognized by fishery managers as a viable resource enhancement technique (McGurrin *et al.*, 1989). Artificial habitats have been deployed throughout the United States in a variety of temperature zones and in fresh, estuarine, and saltwater environments. Their use is perhaps more ubiquitous than in Japan. They are employed for a number of purposes such as recreational and commercial fishing, sport diving, waste disposal, pollution control, and environmental mitigation.

1. History

Despite the lack of quantitative records, an examination of the history of American artificial habitat deployment demonstrates a tradition of enthusiasm and ingenuity in many local communities. The first recorded effort of reef development was the sinking of small log huts to provide habitat and thereby fishing grounds for sheepshead (*Archosargus probatocephalus*,

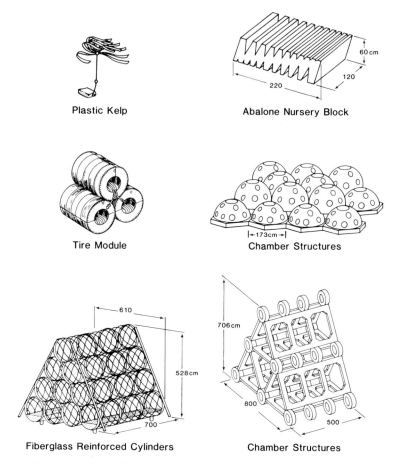

Plastic Kelp   Abalone Nursery Block

Tire Module   Chamber Structures

Fiberglass Reinforced Cylinders   Chamber Structures

Figure 2.5 Representative fabricated structures used in Japanese artificial habitat activities. (A) plastic; (B) abalone nursery block; (C) tire module; (D) chamber structure; (E) fiberglass reinforced cylinders; (F) chamber structure. (Redrawn from Mottet, 1981). (Units in centimeters.)

family Sparidae) in the 1830s in coastal South Carolina (Holbrook, 1860). A century later, major ocean artificial reef construction began in earnest in 1935, further north on the Atlantic coast, with the placement of four vessels and tons of other materials off the New Jersey coast by private interests. Recreational fishing there became so popular that the publicity and increased business that centered around the reef prompted other New Jersey communities to develop more reefs (Stone, 1985a).

Although freshwater artificial reefs were probably in use before the 1930s, it was not until then that researchers and Michigan Conservation Department managers reported on the potential of freshwater habitat enhancement. Hubbs and Eschmeyer (1938) were early advocates of improving lake fishing with brush shelters. Following this initial work, a number of brush shelters were deployed in Michigan lakes and artificial habitat construction became an accepted freshwater fisheries management technique.

Following an inactive period during World War II, there was a resurgence in habitat development in the 1950s. The "Beer-Case Reef" off of New York was typical. There, a brewery donated 14,000 wooden beer cases to a group of charterboat captains who filled them with concrete and sank them in the Atlantic Ocean off of Fire Island.

From the mid-1950s into the 1960s successful reef building efforts were more widely publicized, and numerous recreational organizations on the Atlantic and Gulf of Mexico coasts built small reefs. Many of these efforts were attempted without technical assistance from state or local agencies, were poorly organized, and depended on volunteer labor and donations. They often failed. Many problems (e.g., improper siting and failure of materials) were encountered because of inadequate management and planning, insufficient funding, and an unreliable supply of materials. Lack of communication and exchange of information between the states and organizations involved led to duplication of mistakes and lack of awareness of innovations in technology. Practical solutions to these issues were not readily developed (see Stone, 1985a).

In the late 1950s and throughout the 1960s, research on artificial habitat began. From 1966 through 1974, the U. S. Government operated a marine habitat research program (Stone et al., 1979). Its focus was to determine how reefs could best be used to help develop and conserve recreational fishery resources. The program developed information on construction costs, biology, and management of artificial reefs for state agencies, private organizations, and other reef builders (Parker et al., 1974). After 1974, the federal government reduced its artificial reef activities. Research efforts continued through a variety of state and university studies, although most were descriptive and not quantitative (Steimle and Stone, 1973; Bohnsack and Sutherland, 1985). Until the mid-1980s, reports of research findings typically were provided in various agency documents and not widely circulated.

2. Government Involvement

In the 1970s and 1980s, increased public awareness concerning the decline of certain ocean resources heightened interest in using artificial habitats to rehabilitate or enhance fisheries habitat in the United States (Colunga and Stone, 1974; Aska, 1981; Murray, 1982; D'Itri, 1985). This new momentum

led to passage in the United States of the National Fishing Enhancement Act
of 1984, the development of the U.S. National Artificial Reef Plan in 1985,
and revitalization of a national focus for reef activities, including new research at the federal, state, and university levels.

The National Fishing Enhancement Act stressed the need for responsible and effective efforts to establish artificial reefs in U.S. waters. It defined
an artificial reef as "a structure which is constructed or placed . . . for the
purpose of enhancing fishery resources and commercial and recreational
fishing opportunities."

It also acknowledged the goals of most artificial reef projects:

> . . . properly designed, constructed, and located artificial reefs . . . can enhance
> the habitat and diversity of fishery resources; enhance United States recreational
> and commercial fishing opportunities; increase the production of fishery prod-
> ucts in the United States; increase the energy efficiency of recreational and com-
> mercial fisheries; and contribute to the United States and coastal economies.
> (Appendix C, Stone, 1985b).

These goals are aimed at providing a variety of biological, social, and
economic benefits that have led to increased public interest in artificial habitats. As a direct result of this interest, the Fishing Enhancement Act mandated the development of the National Artificial Reef Plan (Stone, 1985b). It
represents a joint effort by various artificial reef interests including fishermen, divers, researchers, conservation groups, private corporations, and
government agencies. Apparently only the United States and Japan have
published national plans.

The U.S. plan is both a comprehensive presentation of national concerns
on reef development and also a "working document" that serves as a starting
point for effective artificial reef development. The plan serves three major
functions: (1) It provides guidance to individuals, organizations, and agencies
on technical aspects of artificial reef development and management. (2) It is
a technical reference for federal and state agencies involved in meeting standards for reef permitting and management. (3) It encourages the development of systematic regional, state, and local artificial reef plans that focus on
criteria for specific conditions and uses in those areas. Although the plan
provides basic guidance for construction, monitoring, and evaluation of artificial reefs in the U.S.—both freshwater and marine—it does not provide
a funding mechanism, which is organized in a different way. Beyond permitting in federal waters, however, the plan does not mandate central control
of reef-building in the United States.

The most notable source of funds has been the Federal Aid in Sport Fish

Restoration Program. It is a matching grant program with the federal government providing 75% of the funds and individual states supplying the remaining 25% (Christian, 1989). Administered by the U.S. Fish and Wildlife Service (USFWS), these funds result from a 10% excise tax on sport fishing equipment. In fiscal year 1987, more than U.S. $140 million was apportioned from the expanded Sport Fish Restoration Program to states for work on all aspects of sport fish management. Many states will use some of this money for habitat enhancement projects that may include artificial fishing reefs and fish-aggregation devices. In the past, at least 20 inland and coastal states have undertaken habitat enhancement projects using these funds. The recent "Wallop-Breaux amendments" have derived additional funds, from sales of tackle boxes, electric trolling motors, and flasher-type fish finders, from import duties on yachts and pleasure craft, and from taxes on motor boat fuel. These funds are collected at the federal level and apportioned to the state agency responsible for managing sport fishery resources.

An additional program derives from Saltonstall-Kennedy fishery development legislation. A tax on imported fishery products has provided federal money for a number of national and regional artificial reef activities. With assistance from the U.S. National Marine Fisheries Service (NMFS), the private Sport Fishing Institute created the Artificial Reef Development Center in 1983 to assist reef developers with information services, research on practical problems of project development, and public education about reef benefits, limitations, and responsible application of technologies.

The most recent approach to government involvement in artificial habitat planning and funding has been through the regional interstate compacts. In 1986, the Atlantic States Marine Fisheries Commission (ASMFC) formed an Interstate Artificial Reef Program. The program is carried out by a committee of artificial reef program coordinators from the 15 member states of the ASMFC and representatives from NMFS and the USFWS. The goal of the Interstate Artificial Reef Program is to promote effective artificial reef fishery development and provide information to satisfy present and upcoming artificial reef management needs. To meet this goal, the ASMFC has conducted a number of cooperative artificial reef projects including the following: (1) development and assessment of a coastwide data base on reef programs and projects; (2) review of Atlantic coast research needs; (3) establishment of policy and regulatory recommendations for Atlantic coast states; and (4) development and support of legislative proposals for artificial reef programs (see McGurrin et al., 1988; Reeff et al., 1990; Steimle and Figley, 1990). The Gulf States Marine Fisheries Commission recently established a similar program. Cooperative effort in the Great Lakes is described below.

3. State Marine Programs

Concurrent with national emphasis, local plans have been developed by various states. From 1986–1990, the states of New Jersey, North Carolina, New York, California, Louisiana, and Texas developed artificial reef management plans. Thus, six of the 23 coastal states have state artificial reef management plans that deal with local needs and planning efforts. Several other states are considering the development of plans. The National Artificial Reef Plan (Stone, 1985b), Artificial Reef Development for Recreational Fishing: A Planning Guide (Ditton and Burke, 1985), and Profile of Atlantic Artificial Reef Development (McGurrin *et al.*, 1988) have been helpful in this process.

There is no formal worldwide data base to quantitatively describe global development of artificial habitats. However, an initial attempt at organizing a national data base for the United States was successfully conducted in 1984 by the Sport Fishing Institute (SFI), and the following summary is derived from this data.

An assessment of U.S. marine artificial reef activities (McGurrin *et al.*, 1989) was developed from a data base that includes information on the 23 coastal states and documents 572 permitted artificial reef sites in U.S. marine or estuarine waters as of October, 1987. Of the 23 coastal states, 14 have government- (state or local) sponsored marine reef programs. A few states (Delaware and Oregon) have artificial reefs in their waters, but government agencies do not sponsor the reef development. The most active state in the United States by far is Florida (over 200 permitted sites). Florida's situation is most unusual in that many of the reef projects are administered by county government agencies and local groups and are independent of any sort of centralized planning (Seaman and Aska, 1985). The Atlantic and Gulf of Mexico areas are the most active regions (Fig. 2.6).

Approximately 85% of all U.S. marine or estuarine reefs are benthic structures. The remainder are midwater and floating structures, either alone or in combination with other reef types. The majority (80%) of the reef structures in the reef profiles data base are constructed solely of materials of opportunity. The most common materials of opportunity used for marine artificial reefs are ship hulls, surplus concrete, tires, and various types of stone rubble (McGurrin *et al.*, 1989). The largest materials of opportunity are obsolete oil platforms (Fig. 2.7), which may be important to future reef development in the Gulf of Mexico (Reggio, 1989). In contrast to Japan, which has developed some of the most advanced and largest fabricated structures in the world, fabricated habitat materials in the United States have been limited mainly to metal, fiberglass, and nylon midwater FADs and, more recently, small benthic structures made of concrete or plastic (Grove

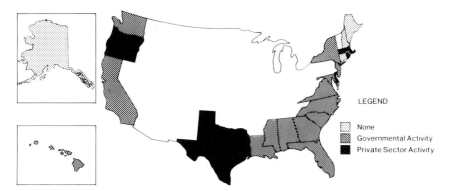

Figure 2.6 U.S. marine artificial habitat programs and activities by state (after McGurrin *et al.*, 1989).

Figure 2.7 The physical scale of structure and aspects of design, construction, and handling are similar for obsolete petroleum structures employed in the United States and steel reefs used in Japan (see Figure 1.4). (Photograph courtesy of U.S. Minerals Management Service.)

and Sonu, 1983). Many habitat materials are used in experimentation. A small number of U.S. businesses are fabricating benthic and midwater structures, and some local or state fishery agencies are using these materials or fabricate their own for studies. In South Carolina, for example, manufactured structures were recently evaluated including concrete pipes, plastic cones and hemispheres, and steel cubes (Bell *et al.*, 1989). In California, natural materials (specifically quarry rock) are used extensively for reef deployment and study.

4. State Freshwater Programs

Techniques employed in the United States to increase freshwater habitat have ranged from the judicious placement of large boulders, Christmas trees (e.g., spruce, fir, pine), tire units, and felled trees, to the construction and placement of elaborate artificial reefs and fish attractors. These structures may involve a multitude of natural and artificial materials designed to improve spawning substrate, increase fish biomass, and attract fish (Prince and Maughan, 1978). Numerous illustrations are provided in Phillips (1990).

No freshwater data base is presently available. However, the SFI Bulletin (1984) summarized a survey of fish habitat development programs in U.S. inland waters. This information was tabulated for all 50 state fishery agencies and could easily be formatted for a computerized data base. Thirty-two state fishery agencies had ongoing artificial habitat installation programs. Approximately U.S. $1.5 million has been spent annually by all state agencies in recent years for construction and installation of freshwater structures. This has resulted in 44,643 individual structures in 1582 bodies of water, including 427 coldwater streams, 45 warmwater streams, and 1110 lakes and reservoirs. Logs, primarily for stream structures (48%), brush (27%), and tires (21%) were most frequently used. Wooden pallets (stake beds), construction rubble, quarry stone, and gabions also were used to a lesser degree. Artificial spawning channels were used in six states, and 18 states have experimental programs to evaluate the effectiveness of specially fabricated polypropylene filament units as fish attractors. Most of the state agencies with ongoing fish attractor installation programs, indicated they planned to continue their programs in the future at current levels (24 states) or at increased levels (seven states). One state indicated a reduction of effort in future years.

Only 18 state agencies (36%) reported that they did not have active development programs in place. However, four had installed fish habitats in a few bodies of water for experimental evaluation. Another agency only provides design and construction advice for local sportsmen's club projects. In two other states, federal agencies (U.S. Forest Service and U.S. Bureau of Land Management) conduct projects on federally administered lands.

It is important to note that the SFI survey was completed prior to the expansion of the Federal Aid in Sport Fish Restoration revenues. Since then, several states have expanded their efforts. As is the case for marine reefs, most of these efforts are small-scale projects conducted by state agencies for the improvement of local recreational fishing opportunities. However, the Great Lakes region is the site of a number of new large-scale reef projects. There, the Habitat Advisory Board of the Great Lakes Fishery Commission appointed an artificial reef task force and approved a position statement and evaluation guidelines for artificial reefs (Gannon, 1990). The report recommends that artificial reefs be considered experimental since their use as a fishery management technique is new in this region. Consistent with themes expressed elsewhere in this volume, the report urges that reefs be constructed only to achieve fishery management objectives and not as a pretext for solid waste disposal. Further evaluation of ecological and socioeconomic performance on a long-term basis is a recommendation that applies not only to the Great Lakes, but to all habitats where alien materials and structures are added.

Collectively, the approaches of Japan and the United States offer a significant source of information for planning or implementing artificial habitat projects and programs to meet different fishery needs and management objectives. In concert with findings from other parts of the world, this knowledge base can serve as the foundation for periodic handbooks and summaries of practices that incorporate new research and technologies.

References

Aprieto, V. L. 1988. Aspects of management of artificial habitats for fisheries: The Philippine situation. Pp. 102–110 in Report of the Workshop on Artificial Reefs Development and Management, ASEAN/SF/88/GEN/8. Penang, Malaysia.

Ardizzone, G. D., M. F. Gravina, and A. Belluscio. 1989. Temporal development of epibenthic communities in the central Mediterranean Sea. Bulletin of Marine Science 44:592–608.

Aska, D. Y., editor. 1981. Artificial reefs: Conference proceedings, Report 41. Florida Sea Grant College, Gainesville.

Bell, M., C. J. Moore, and S. W. Murphey. 1989. Utilization of manufactured reef structures in South Carolina's marine artificial reef program. Bulletin of Marine Science 44:818–830.

Bellan-Santini, D. 1982. Exposé des travaux de la réunion. Pp. 9–11 in Journée d'études sur les aspects scientifiques concernant les récifs artificiels et la mariculture suspendue. Commission Internationale pour l'Exploration Scientifique de la Mer Méditerranée, Monaco.

Bohnsack, J. A., and D. L. Sutherland. 1985. Artificial reef research: A review with recommendations for future priorities. Bulletin of Marine Science 37:11–39.

Bombace, G. 1989. Artificial reefs in the Mediterranean Sea. Bulletin of Marine Science 44:1023–1032.

Bortone, S. A., R. L. Shipp, W. P. Davis, and R. D. Nester. 1988. Artificial reef development along the Atlantic coast of Guatemala. Northeast Gulf Science 10:45–48.

Buckley, R. M., J. Grant, and J. Stephens, Jr. 1985. Third International Artificial Reef Conference 3–5 November 1983, Newport Beach, California. Bulletin of Marine Science 37:1–2.

Buckley, R. M., D. G. Itano, and T. W. Buckley. 1989. Fish aggregation device (FAD) enhancement of offshore fisheries in American Samoa. Bulletin of Marine Science 44:942–949.

Campos, J. A., and C. Gamboa. 1989. An artificial tire-reef in a tropical marine system: A management tool. Bulletin of Marine Science 44:757–766.

Chang, K., and K. Shao. 1988. The sea farming projects in Taiwan. Acta Oceanographica Taiwanica 19:52–59.

Christian, R. T. 1989. Dingell-Johnson/Wallop-Breaux: The Federal Aid in Sport Fish Restoration Program Handbook, 2nd edition. Sport Fishing Institute, Washington, D.C.

Colunga, L. C., and R. B. Stone, editors. 1974. Proceedings of an International Conference on Artificial Reefs, TAMU-SG-74-103. Texas A&M University Sea Grant College Program, College Station.

De Bernardi, E. 1989. The Monaco Underwater Reserve—design and construction of artificial reefs. Bulletin of Marine Science 44:1066.

D'Itri, F. M., editor. 1985. Artificial reefs: Marine and freshwater applications. Lewis Publishers, Inc., Chelsea, Michigan.

Ditton, R. B., and L. B. Burke. 1985. Artificial reef development for recreational fishing: A planning guide. Sport Fishing Institute, Artificial Reef Development Center, Washington, D.C.

Downing, N., R. A. Tubb, C. R. El-Zahr, and R. E. McClure. 1985. Artificial reefs in Kuwait, northern Arabian Gulf. Bulletin of Marine Science 37:157–178.

Duff, D. A., and M. Gnehm. 1986. Indexed bibliography on stream improvement. U.S. Department of Agriculture Forest Service, Logan, Utah.

Eckert, R. D. 1979. The enclosure of ocean resources. Hoover Institution Press, Stanford, California.

Fabi, G., L. Fiorentini, and S. Giannini. 1989. Experimental shellfish culture on an artificial reef in the Adriatic Sea. Bulletin of Marine Science 44:923–933.

Gannon, J. E., editor. 1990. International position statement and evaluation guidelines for artificial reefs in the Great Lakes, Special Publication 90-2. Great Lakes Fishery Commission, Ann Arbor, Michigan.

Gates, P. D. 1990. Review of Pacific Island FAD deployment programs. 20 pp. (unnumbered) *in* Workshop on Fish Aggregating Devices (FADs), 22nd Regional Technical Meeting on Fisheries. South Pacific Commission, Noumea, New Caledonia.

Grove, R.S., and C. J. Sonu. 1983. Review of Japanese fishing reef technology, Technical Report 83-RD-137. Southern California Edison Company, Rosemead, California.

Grove, R. S., C. J. Sonu, and M. Nakamura. 1989. Recent Japanese trends in fishing reef design and planning. Bulletin of Marine Science 44:984–996.

Holbrook, J. E. 1860. Ichthyology of South Carolina, 2nd edition. John Russell, Charleston, South Carolina.

Hubbs, C. L., and R. W. Eschmeyer. 1938. The improvement of lakes for fishing: A method of fish management, Bulletin 2. Michigan Department of Conservation, Institute of Fisheries Research, Ann Arbor.

Ino, T. 1974. Historical review of artificial reef activities in Japan. Pp. 21–23 *in* L. Colunga and R. B. Stone, editors. Proceedings of an International Conference on Artificial Reefs, TAMU-SG-74-103. Texas A&M University Sea Grant College Program, College Station.

Japan Coastal Fisheries Promotion Association (JCFPA). 1984. Coastal fisheries development program structural design guide. (In Japanese)

Japan Coastal Fisheries Promotion Association (JCFPA). 1986. Artificial reef fishing grounds construction planning guide. (In Japanese)

Kapetsky, J. M. 1981. Some considerations for the management of coastal lagoon and estuarine fisheries, FAO Fisheries Technical Paper 218. Food and Agricultural Organization of the United Nations, Rome.

McGurrin, J. M., and ASMFC Artificial Reef Committee. 1988. A profile of Atlantic artificial reef development, Special Report 14. Atlantic States Marine Fisheries Commission, Washington, D.C.

McGurrin, J. M., R. B. Stone, and R. J. Sousa. 1989. Profiling United States artificial reef development. Bulletin of Marine Science 44:1004–1013.

Mottet, M. G. 1981. Enhancement of the marine environment for fisheries and aquaculture in Japan, Technical Report 69. Washington Department of Fisheries, Olympia.

Murray, J. D., editor. 1982. Mid-Atlantic Artificial Reef Conference: A collection of abstracts, NJSG-82-78. Marine Sciences Consortium, State University of New Jersey, New Brunswick.

Naeem, A. 1989. Fish aggregating devices (FADs) in the Maldives. Ministry of Fisheries, Republic of Maldives, Male.

Parker, R. O., R. B. Stone, C. C. Buchanan, and F. W. Steimle. 1974. How to build marine artificial reefs, Fishery Facts Number 10. National Marine Fisheries Service, National Oceanic and Atmospheric Administration, Washington, D.C.

Phillips, S. H. 1990. A guide to the construction of freshwater artificial reefs. Sport Fishing Institute, Washington, D.C.

Picken, G. B., and A. D. McIntyre. 1989. Rigs to reefs in the North Sea. Bulletin of Marine Science 44:782–788.

Pollard, D. A., and J. Matthews. 1985. Experience in the construction and siting of artificial reefs and fish aggregation devices in Australian waters, with notes on and a bibliography of Australian studies. Bulletin of Marine Science 37:299–304.

Pramokchutima, S., and S. Vadhanakul. 1987. The use of artificial reefs as a tool for fisheries management in Thailand. Pp. 427–441 in Indo-Pacific Fishery Commission. Symposium on the Exploitation and Management of Marine Fishery Resources in Southeast Asia, RAPA Report 1987/10. Regional Office for Asia and the Pacific, Food and Agriculture Organization of the United Nations, Bangkok.

Preston, G. L. 1990. Fish aggregation devices in the Pacific Island region. 21 pp. (unnumbered) in Workshop on Fish Aggregating Devices (FADs), 22nd Regional Technical Meeting on Fisheries. South Pacific Commission, Noumea, New Caledonia.

Prince, E., and O. E. Maughan, 1978. Freshwater artificial reefs: Biology and economics. Fisheries 3:5–9.

Reeff, M. J., and J. M. McGurrin. 1986. The artificial reef source: an annotated bibliography of scientific, technical, and popular literature. Sport Fishing Institute, Artificial Reef Development Center, Washington, D.C.

Reeff, M. J., J. M. Murray, and J. McGurrin. 1990. Recommendations for Atlantic state artificial reef management, Recreational Fisheries Report 6. Atlantic States Marine Fisheries Commission, Washington, D.C.

Reggio, V. C., Jr., compiler. 1989. Petroleum structures as artificial reefs: A compendium. Fourth International Conference on Artificial Habitats for Fisheries, Rigs-to-Reefs Special Session, November 4, 1987, OCS Study/MMS 89-0021. U.S. Minerals Management Service, New Orleans, Louisiana.

Relini, G., and L. O. Relini. 1989. Artificial reefs in the Ligurian Sea (northwestern Mediterranean): Aims and results. Bulletin of Marine Science 44:743–751.

Sanjeeva Raj, P. J. 1989. Modified artisanal artificial fish habitats on the Tamil Nadu coast of India. Bulletin of Marine Science 44:1069–1070.

Seaman, W., Jr. In press. Fourth International Conference on Artificial Habitats for Fisheries: A convener's perspective. Proceedings of 40th Annual Conference, Gulf and Caribbean Fisheries Institute.

Seaman, W., Jr., and D. Y. Aska. 1985. The Florida reef network: Strategies to enhance user benefits. Pp. 545–561 *in* F. M. D'Itri, editor. Artificial reefs: Marine and freshwater applications. Lewis Publishers, Inc. Chelsea, Michigan.

Seaman, W., Jr., R. M. Buckley, and J. J. Polovina. 1989. Advances in knowledge and priorities for research, technology and management related to artificial aquatic habitats. Bulletin of Marine Science 44:527–532.

Shao, K.-T. 1988. A master plan of artificial reef project along the northern coast of Taiwan, Monograph Series No. 12. Institute of Zoology, Academia Sinica, Taipei.

Spanier, E., M. Tom, and S. Pisanty. 1985. Enhancement of fish recruitment by artificial enrichment of man-made reefs in the southeastern Mediterranean. Bulletin of Marine Science 37:356–363.

Sport Fishing Institute. 1984. Freshwater fish attractors. SFI Bulletin, September 1984, No. 359:6–7.

Stanton, G., D. Wilber, and A. Murray. 1985. Annotated bibliography of artificial reef research and management, Report 74. Florida Sea Grant College, Gainesville.

Steimle, F., and W. Figley. 1990. A review of artificial reef research needs, Recreational Fisheries Report 7. Atlantic States Marine Fisheries Commission, Washington, D.C.

Steimle, F., and R. B. Stone. 1973. Bibliography on artificial reefs. Coastal Plains Center for Marine Development Services, Wilmington, North Carolina.

Stone, R. B. 1985a. History of artificial reef use in the U.S. Pp. 3–9 *in* F. M. D'Itri, editor. Artificial reefs: Marine and freshwater application. Lewis Publishers, Inc., Chelsea, Michigan.

Stone, R. B., compiler. 1985b. National artificial reef plan, NOAA Technical Memorandum NMFS OF-6. U.S. National Marine Fisheries Service, Washington, D.C.

Stone, R. B., H. L. Pratt, R. O. Parker, Jr., and G. E. Davis. 1979. A comparison of fish populations on an artificial and natural reef in the Florida Keys. Marine Fisheries Review 41:1–11.

U.S. National Research Council. 1988. Fisheries technology for developing countries. Report of an ad hoc Panel of the Board on Science and Technology for International Development. National Academy Press, Washington, D.C.

Welcomme, R. L. 1972. An evaluation of the acadja method of fishing as practised in the coastal lagoons of Dahomey (West Africa). Journal of Fish Biology 4:39–55.

Yamane, T. 1989. Status and future plans of artificial reef projects in Japan. Bulletin of Marine Science 44:1038-1040.

3

Ecology of Artificial Reef Habitats
and Fishes

J. A. BOHNSACK
Southeast Fisheries Center
National Marine Fisheries Service
Miami, Florida

D. L. JOHNSON
School of Natural Resources
Ohio State University
Columbus, Ohio

R. F. AMBROSE
Marine Science Institute
University of California, Santa Barbara
Santa Barbara, California

I. Introduction

Artificial reefs and other aquatic habitats created from natural materials and man-made structures offer a potential opportunity for improving habitat, increasing resources, and manipulating assemblages of organisms in ways that benefit humankind (Buckley *et al.*, 1985; Seaman *et al.*, 1989). Ecological factors that operate on natural reefs, such as physical disturbance, recruitment, competition, and predation, also will operate on so-called artificial habitats. In theory, the ecology of artificial reefs should be no different than that of natural reefs except for differences caused by the design and positioning of artificial structures. Many species that utilize natural reefs have morphological, behavioral, and physiological adaptations that preadapt them to successfully exploit artificial habitats in freshwater and saltwater environments.

Lack of knowledge concerning ecology of artificial habitats is a central

Artificial Habitats for Marine and Freshwater Fisheries
Copyright © 1991 by Academic Press, Inc.
All rights of reproduction in any form reserved.

problem in the debate on their proper use in fishery management (Stone, 1985; Buckley, 1989; Meier *et al.*, 1989; Polovina, 1989), and also restricts artificial habitat application as mitigation tools for various environmental damages. Better understanding will allow better design and more effective use of these structures. It also may answer questions about the worth of building reefs under different circumstances.

This chapter provides a general review of artificial habitat ecology and gives examples of how specific local conditions influence patterns of species distribution and abundance. Special emphasis is on fishes. Although in one section the ecology of marine and freshwater artificial structures is contrasted, elsewhere their ecologies are treated together under appropriate chapter headings. The principles of ecology are assumed to be similar in all aquatic systems, therefore the same biotic and abiotic factors must be taken into consideration when designing and placing habitat structure, regardless of location.

For present purposes the term artificial reef is used primarily to describe benthic structures created accidentally or deliberately by human activities. The more general term, artificial habitat, refers to structures deployed either on or above the sea floor, including floating or midwater fish-aggregating devices (FADs). (In some cases, it is not possible to discern if literature accounts are focused on benthic or nonbenthic structures.) Emphasis is given to recently published studies. Whereas our intent is to provide global coverage, the reader may recognize a bias toward literature from North America and tropical and subtropical marine areas. In part this reflects the available research record, as described in Chapter 1.

Artificial habitat ecology is a young, rapidly growing science with most studies having taken place during the past two decades. Although many generalizations concerning artificial habitat ecology have been made (e.g., Bohnsack and Sutherland, 1985), they should be taken cautiously. Generalizations often have been extrapolated from few data or very restricted studies. Because of the numbers of organisms and the complexity of their interactions, many artificial habitat studies have focused on specific groups (e.g., algae, fishes, invertebrates), processes such as competition or colonization, or particular guilds, including fouling organisms.

The variety of materials used for artificial habitats and the variety of conditions in which they are deployed also make generalizations difficult or of questionable validity. In the ocean, materials employed range from accidental shipwrecks (Fig. 3.1), to discarded scrap materials haphazardly deposited, to carefully designed and fabricated structures (Fig. 3.2) deployed in specific patterns. Some structures designed for other purposes serve as artificial reefs, including shore protection structures (Hay and Sutherland, 1988) and oil and gas production platforms (Reggio, 1989). (See Chapter 4

Figure 3.1 Shipwrecks, although accidental, often provide effective artificial habitat for a variety of organisms.

Figure 3.2 Fabricated concrete reef modules deployed near Miami, Florida.

for a review of engineering practices and Figure 1.1 in Chapter 1 for illustrations of principal types of structures.)

Artificial habitats have been used in freshwater and marine environments with water conditions ranging from shallow to deep; tropical to temperate; clear to turbid; with weak to strong currents or zero to high turbulence. Regionally, they may differ markedly in location and composition, perhaps due to availability of materials or variation of some underlying ecological factor. For marine reefs in the United States, for example, fewer habitats are constructed at higher latitudes. Off of California, 33 of 37 constructed sites were south of 34°21′ N. (approximately Point Conception), and on the Atlantic coast 216 sites (79%) were in southeastern states (McGurrin *et al.*, 1989b). According to data in McGurrin *et al.* (1988, 1989a) and Ambrose *et al.* (1989) quarry rock is the most common item on California reefs (at 23 sites) but little used on the Atlantic coast, while vessels are much more common in the eastern United States (199 of 273 sites).

The purposes for which artificial habitat structures have been used vary greatly, as well. For example, in the largest programs the emphasis in Japan is on fabricated structures for commercial fishing reefs, whereas in the United States almost all habitats have been built of scrap ("materials of opportunity") primarily to enhance recreational fishing (see Chapter 2). The interactions of diverse management goals, materials used and different locations for deployment, and how they affect artificial reef ecology, are poorly known, primarily because of a lack of comparable controlled scientific studies.

II. Assemblage Structure

Perhaps one of the major problems in understanding, and certainly in reviewing and making generalizations about artificial habitat ecology, is the daunting task of dealing with myriads of organisms that utilize artificial habitats. Also, species vary geographically and in time. Sampling is a major problem in many cases, as discussed in Chapter 6. To simplify this complexity, researchers usually classify organisms into convenient categories. Often the conclusions derived are biased or limited by the classification used. Some of the major classification approaches used in artificial habitat literature, with their advantages and especially their disadvantages, are discussed below.

A. Components

Biotic components of artificial habitat assemblages are usually classified based on taxonomy, body size, mode of existence, or trophic level. Many studies, for example, are restricted to algae, fouling assemblages (periphyton in freshwater), invertebrates, macroinvertebrates, or fishes, depending on the expertise or interests of the researchers. Frequently studies are restricted to a single or few species, often only those species with direct commercial or recreational importance.

This chapter refers to organisms at an artificial habitat as an "assemblage" rather than a community, because the term community implies a coevolved, deterministic system that has emergent properties (Mapstone and Fowler, 1988). Assemblage is a more neutral term.

Understanding the dynamics and ecology of one component of an artificial reef assemblage may be impossible without a comprehensive knowledge of other components (Werner, 1986). For example, composition and competition between species of fouling assemblages may vary greatly depending on fish grazing pressure (Hixon and Brostoff, 1985). The abundance and distribution of economically important fishes also may be determined by other factors such as food availability, competition with other species, or overall habitat productivity. This may not be otherwise understood without comprehensive ecological research (Werner, 1986).

Any method of classifying assemblages produces its own biases. For example, a simple taxonomic listing of fishes that greatly change body size with growth may mask drastic ontogenetic (age-dependent) niche shifts caused by changes in ecology or morphology of individuals. Ontogenetic changes may be continuous or abrupt depending on the species (Werner, 1986).

Many species have different ecological roles depending on their size and the stage of their life cycle. Changes in body size may influence resource utilization abilities, especially by increasing the type and range of food sizes (Werner, 1986), and tend to reduce risks of predation (Menge and Sutherland, 1976, 1987). Fishes have been shown to stay near artificial structures for protection when small, but when larger and less vulnerable to predation, they spend more time away from habitats in marine environments (Anderson et al., 1989) and freshwater (Werner, 1986).

A common and often unacknowledged bias occurs when making evaluations based solely on the number of individuals of the species present. This gives great emphasis to small individuals while ignoring other important assemblage measures such as individual size, biomass, function, and frequency-of-occurrence. For example, tomtate (*Haemulon aurolineatum*) comprised 74% of numbers of individuals but only 10% of the total fish biomass in one study (Bohnsack et al., 1989). On artificial reefs off of southern California (Anderson et al., 1989) and southeastern Florida (Bohnsack et al., 1989) in the United States, numerical abundance varied greatly due to variable recruitment episodes, although biomass density was less variable over time. DeMartini et al. (1989) described biases caused by measuring density when total abundance is actually the variable of interest.

The mode of existence of individual organisms is often used to classify assemblages. Attached biota, referred to as fouling organisms in marine systems, include algae, sponges, anemones, barnacles, and tunicates, among other groups. In freshwater, attached algae is referred to as periphyton, and associated invertebrates include bryozoans, sponges, hydra, and mussels. Free-living organisms include most crustaceans and fishes that may be classified in terms of their spatial positioning relative to a habitat. Nakamura (1985) distinguished fishes that are usually in contact with the reef and tend to live in holes, fishes that tend to swim around a reef while remaining near the bottom, and fishes that hover in midwater above a reef. (See Chapter 4, Figure 4.11.) These responses vary by species and their stage of maturation.

Trophic classifications are based on the means of obtaining food. Food resources can influence the abundance of various components of an assemblage. Major food items include algae, invertebrates, and fishes at an artificial habitat, but also passing plankton and organisms in accessible surrounding areas (Randall, 1963; Hueckel and Stayton, 1982; Prince et al., 1985). Trophic levels may change with age or size, often due to morphological or behavioral changes (Werner, 1986). Largemouth bass (*Micropterus salmoides*), for example, initially feed on zooplankton but later switch to littoral invertebrates and then to other fishes as adults (Gilliam, 1982). Young bluegill (*Lepomis macrochirus*) are initially pelagic schoolers but become more structure-oriented as juveniles and adults. Similar ontogenetic, trophic, and habitat shifts have been noted for many marine species (Shulman and Ogden, 1987; Bohnsack, 1989). Although assemblages are often variable, carnivores tend to dominate the biomass on marine artificial habitats. Brock and Grace (1987), for example, found that carnivores dominated the fish biomass (average 70%; range 40% to 90%) followed by herbivores, planktivores, and omnivores on a Hawaiian artificial habitat.

Unattached organisms show various residency patterns. Species present may range from opportunists that briefly utilize artificial reefs to obligatory species that critically depend on a reef structure for part of their lives. Species can also be classified as residents, visitors, or transients (Talbot et al., 1978; Bohnsack and Talbot, 1980). Residents tend to stay at a structure for long periods once they colonize or settle. Visitors use artificial habitats for brief periods, from a few minutes, hours, days, or seasonally. Transients are species observed at artificial habitats but that do not respond to it differentially from surrounding structures. Because of differences in individual behavior, residency determinations are sometimes subjective, unless tagging studies are conducted.

Similar residency composition has been reported in some studies. In a three-year comparison of six U.S. (U) and eight Australian (A) tropical reefs, Bohnsack and Talbot (1980) found similar fish assemblage patterns in terms of total families (31 A, 31 U); total observed species (88 A, 89 U); primary residents (68% A, 62% U); visitors (22% A, 26% U); transients (6% A, 8% U); and mean number of individuals per reef (87 A, 65 U). The mean number of species per reef, however, was significantly greater in the United States (16.6) versus Australia (12.8). Bohnsack *et al.* (1989) reported similar residency patterns (72% residents, 17% visitors, and 10% transients) based on 127 species observed on larger fabricated reefs off southeastern Florida.

Transients and visitor species have been ignored in some studies. These omissions could lead to misleading interpretations. Visitors to fabricated concrete artificial reefs off southeastern Florida comprised only 1.5% of the total number of fishes censused, but they represented 38.8% of the total fish biomass and were dominated by species of economic value (Bohnsack *et al.*, 1989). In the same study, some transient and visiting species, although present only briefly, were potentially important influences as predators or competitors. For example, transient mackerel scad (*Decapterus macarellus*) represented considerable biomass and could have potential ecological importance as predators on fish eggs and larvae (affecting recruitment) and as competitors for planktonic food resources.

Behavior of a species may differ between reefs. For example, a species may be a resident on a large reef but only a visitor to a small structure because there are insufficient resources (food or shelter) to support a permanent population. Ogden and Buckman (1973) noted striped parrotfishes (Scaridae) were absent on small natural reefs apparently due to insufficient food resources. Also, fishes may be more likely to travel between habitats that are close together (see Bohnsack and Sutherland, 1985; Bohnsack, 1989).

B. Comparison of Artificial and Natural Habitat Assemblages

Recent reviews on the ecology of natural and artificial habitats can be found for fouling organisms (Cairns, 1982), fishes (Sale, 1980; Werner, 1986; Lowe-McConnel, 1987), and artificial habitats (Buckley *et al.*, 1985; D'Itri, 1985; Seaman *et al.*, 1989).

The simplest ecological evaluation of artificial habitats is a mere taxonomic listing of species observed. By itself, this is rarely much help in understanding artificial habitat ecology because abundances vary greatly, and almost any species that occurs associated with natural habitats can usually be found at artificial structures (e.g., Ambrose and Swarbrick, 1989).

One approach to artificial reef ecology asks how artificial reef assemblages differ from those on natural reefs by testing null hypotheses of no difference between artificial habitat assemblages and nearby natural habitat communities for various parameters and functions. Failure to reject null hypotheses implies that artificial habitats function like natural habitats, and their ecological contribution is directly proportional to the amount of bottom covered.

Most comparative studies of artificial and natural reef assemblages in the same area show great similarity in species composition, although species abundance and biomass may differ considerably (Ambrose and Swarbrick, 1989; Bohnsack and Sutherland, 1985; Matthews, 1985; Solonsky, 1985). Ambrose and Swarbrick (1989) and DeMartini *et al.* (1989) found higher densities of fishes on artificial reefs than nearby natural reefs. Ratios of catch rates or fish standing stock between artificial and natural habitat (reef and other natural bottom habitat) for freshwater and saltwater are depicted in Figs. 3.3 and 3.4, respectively. Results show higher fish catches and density at artificial habitats, although the data might be biased because artificial reef failures are less likely to be reported in the literature. For example, information has not been published from Gerber's (1987) thesis that shows no significant differences in catch-per-effort by gill nets or controlled angling on rock reefs added to Lake Erie, as compared to reference areas. Little

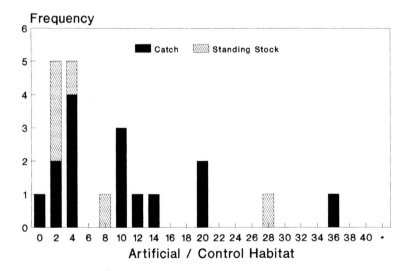

Figure 3.3 Representative ratios of catches or standing stock at North American artificial versus control habitats in freshwater. (Reported units vary. Data sources: Burress, 1961; Crumpton and Wilbur, 1974; Gannon *et al.*, 1985; Patriarche, 1959; Pettit, 1973; Pierce and Hooper, 1980; Prince *et al.*, 1985.)

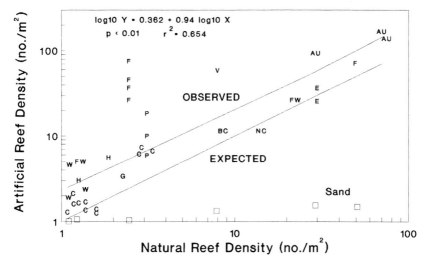

Figure 3.4 Regression of marine fish density at artificial reefs versus natural control reefs. Squares show density on sand bottom. The expected line assumes the same density at artificial and natural habitats. Data points represent individual or mean reported values, or values calculated from reported results. In some cases, natural and artificial habitat data were taken from different sources. AU, Australia (Talbot *et al.*, 1978; Bohnsack and Talbot, 1980); BC, British Columbia (Gascon and Miller, 1981); C, California (Matthews, 1985); E, Enewetak (Nolan, 1975); F, southeast Florida (Stone *et al.*, 1979); FW, Florida west coast (Smith *et. al.*, 1979); G, Guam (Kock, 1982); H, Hawaii (McVey, 1970; Brock and Grace, 1987); K, Florida Keys (Bohnsack, 1979); NC, North Carolina (Lindquist *et al.*, 1989); P, Puerto Rico (Fast and Pagan, 1974; R. Bejarano, Louisiana Department of Wildlife and Fisheries); V, Virgin Islands (Shulman, 1984); W, Washington (Walton, 1979; Hueckel *et al.*, 1989).

information exists on other effects of artificial habitat on organisms, although Prince *et al.* (1985) reported that sunfishes (family Centrarchidae) caught at artificial habitats showed better body condition and faster growth in weight than fishes caught elsewhere.

Comparison of worldwide data showed consistently higher fish density ($n = 29$ sites) and fish biomass ($n = 23$) at artificial reefs than on nearby control, natural reefs (Bohnsack, 1991). This analysis is expanded with additional data in Fig. 3.5. Although artificial reefs have higher values than the nearby natural reefs, there appears to be a limit based on the productivity of the region. Areas with few fishes or low biomass on natural reefs would also tend to have corresponding low values on artificial reefs. Areas with abundant fishes or high biomass on natural reefs would tend to have high values on artificial reefs. On the other hand, in large freshwater oligo-

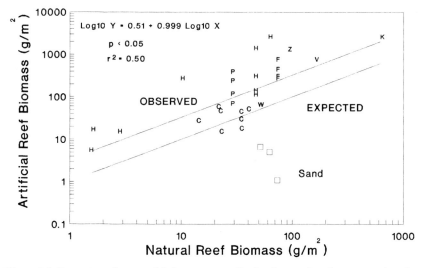

Figure 3.5 Regression of marine fish biomass at artificial and natural reefs. Squares show biomass density on sand bottom. Data were transformed by $(x + 1)$. The expected line assumes the same biomass at artificial and natural habitats. Data points represent individual or mean reported values, or values calculated from reported results. In some cases, natural and artificial habitat data were taken from different sources. C, California (Ambrose and Swarbrick, 1989; DeMartini *et al.*, 1989); F, southeast Florida (Bohnsack, *et. al.*, 1989); H, Hawaii (McVey, 1970; Kanayama and Onizuka, 1973; Brock and Grace, 1987; Brock and Norris, 1989); K, Florida Keys (Bohnsack, 1989); P, Puerto Rico (Fast and Pagan, 1974; R. Bejarano, Louisiana Department of Wildlife and Fisheries); V, Virgin Islands (Randall, 1963); W, Washington (Walton, 1979); Z, New Zealand (Russell, 1975).

trophic lakes, researchers have found it difficult to document increased numbers of fishes at artificial reefs as compared to nonreef areas, based on our observations.

III. Environmental Factors

Artificial habitats can provide sources of food, shelter, and sites for orientation and reproduction. Presence of any species depends on suitable living conditions, a supply of recruits, and higher recruitment and immigration than mortality and emigration. Suitable living conditions include access to food resources, shelter from predators, and normal physical conditions within biological tolerances of an organism. In tropical areas, some evidence suggests that shelter from predation may be more important than food for determining the abundance of fishes (Sale, 1980; Shulman, 1984; Hixon and Beets, 1989; Bohnsack, 1991).

Species absent from artificial habitats but found at similar natural habitats in the same area may (1) lack critical resources; (2) not yet have colonized due to chance; (3) be absent due to nomadic movement patterns or seasonal conditions; (4) be absent due to excessive mortality from predation or environmental stresses; or (5) be absent due to chance (stochastic) factors. Chance factors are more likely to be important for small populations (MacArthur and Wilson, 1967).

A lack of critical resources is the reason for the absence of many species. Tropical boring organisms, for example, may not be found at steel artificial reefs because of the lack of suitable substrate. They may eventually colonize, however, as the reef "ages" and suitable substrate becomes available from fouling processes. Hixon and Beets (1989) showed that larger predators tended not to colonize artificial reefs without suitably sized holes present. Herbivores were notably absent on small artificial and natural reefs probably because of insufficient food resources (Ogden and Buckman, 1973; Bohnsack et al., 1989).

Environmental conditions recognized as important to artificial habitat will be described. These include surrounding substrate type and quantity; isolation from similar habitats; depth; latitude, seasonality and temperature; water quality (salinity, turbidity, pollution); current and surge conditions; and primary productivity.

A. Location

Geographic location of an artificial habitat profoundly affects its ecological features and is probably more important than its physical shape (Bohnsack, 1991). Comparative studies between artificial habitat assemblages tend to indicate high similarity within regions but substantial differences between regions (Bohnsack and Talbot, 1980; Walsh, 1985; Ambrose and Swarbrick, 1989). Potential limits on the population abundance and composition of artificial habitat assemblages are set by environmental conditions.

B. Surrounding Substrate

The type and quantity of substrate surrounding an artificial habitat can influence assemblages by changing the quality and quantity of food resources for fishes (e.g., Davis et al., 1982). Randall (1963) concluded that the presence of seagrass-beds greatly increased the number and biomass of fishes on an experimental artificial reef in the Virgin Islands. The surrounding substrate also may act as an important intermediate habitat that influences recruitment. If it is rock, for example, it might influence algal development and, consequently, fish recruitment (e.g., Wilson and Schlotterbeck, 1989).

C. Isolation

Isolation from similar habitats affects artificial reef assemblages by influencing the number and type of potential recruits and the colonization rate. Shulman (1985) found higher fish recruitment at artificial reef sites away from a large reef due to differences in available shelter and predator encounter rates. Many studies have concluded that the same structure will have different assemblages when placed at different distances from similar habitat (Chang *et al.*, 1977; Molles, 1978; Bohnsack, 1979, 1989, 1991; Gascon and Miller, 1981; Alevizon *et al.*, 1985; Carter *et al.*, 1985; Shulman, 1985; Walsh, 1985; Schroeder, 1987; Wendt *et al.*, 1989). The dynamic implications of isolation are discussed later under island biogeographic theory.

D. Depth

The distributions of many species are depth-related. Major changes in marine fish assemblages have been noted around depths of 30 m off Taiwan (Chang, 1985) and 45 m off Florida (Shinn and Wicklund, 1989). In freshwater, Lynch and Johnson (1983) reported that although adult bluegill were abundant at artificial structures in an Ohio reservoir at 4 m depth, they were almost completely lacking at structures 2 m deep.

Depth alters light conditions and thus influences algal composition, abundance, and productivity. Deeper reefs may lack or have fewer algae (e.g., Mathews, 1983).

Depth also affects temperature; usually, warm water floats above cold water creating a thermocline. Thermoclines can be a barrier to some organisms. In many productive freshwater lakes, thermocline depth is critical because the hypolimnion, the colder layer below the thermocline, is frequently anoxic. Artificial reef assemblages can be influenced because of seasonal changes in thermocline depth especially in temperate freshwater lakes. Lynch and Johnson (1983) found that white crappie (*Pomoxis annularis*) were abundant at structures in 4 m of water until midsummer when the lowering thermocline allowed use of deeper 7 m structures, at which time the shallower structures were abandoned. Prince *et al.* (1985) found that some temperate freshwater reefs at depths of 6 m or less were virtually devoid of fishes in the winter.

E. Latitude, Seasonality, and Temperature

Latitude, seasonality, and temperature are closely related. Seasonal changes affect availability of colonists, which can affect the fouling assemblage (e.g., Woodhead and Jacobson, 1985; Bailey-Brock, 1989) and fish assemblage (e.g., Kock, 1982; Alevizon et al., 1985; Woodhead et al., 1985; Bohnsack, 1991). Extreme hot or cold temperatures can cause mortality depending on specific tolerances of each species (e.g., Bohnsack, 1983b). Also, many fishes are adapted to seasonal changes and shift habitats or areas with season (e.g., Chandler et al., 1985). Prince et al. (1985) showed that some temperate freshwater fishes used artificial structure seasonally when water temperatures exceeded 10° C.

Fewer total numbers of species using temperate artificial reefs is one of the most obvious differences between tropical and temperate marine reefs. Schoener (1982) noted, based on few observations, that the increase in the number of sessile (attached) species accumulating with time is inversely related to latitude. One simple explanation for this difference is that fewer species have evolved at higher latitudes. Gascon and Miller (1981) suggested that temperate artificial reef fish assemblages may be more predictable than those on tropical reefs partly because of a smaller species pool available to colonize temperate reefs.

Based on ecological theory, tropical environments may be less seasonal and more dominated by biological interactions, whereas temperate environments are more seasonal and dominated by physical factors (Menge and Sutherland, 1976, 1987). Holbrook et al. (1990) found only weak spatial variation in fish assemblages on reefs in southern California, which they attributed in part to temporal variation in kelp and algal communities. They suggested that the degree of spatial differentiation in substrate-oriented fish assemblages may be inversely related to the temporal constancy of biogenic reef structures. Thus, some differences between tropical and temperate fish assemblages may be due to differences in habitat constancy. Clearly, more comparative research is needed before the causes and differences between tropical and temperate artificial reefs are fully understood. Studies on variability of recruitment with latitude may be especially useful.

F. Water Conditions

1. Water Quality

Various water quality parameters including salinity, turbidity, and pollution levels may affect assemblages depending on the tolerance of individual species. Relini and Relini (1989), for example, noted failures of artificial reefs to produce fishes due partly to high turbidity and pollution in the

Mediterranean Sea. In contrast, Los Angeles harbor on the U.S. Pacific coast is also considered polluted and turbid, yet artificial structures there have a dense and diverse fish fauna (Ambrose and Swarbrick, 1989) and flourishing populations of giant kelp (Rice *et al.*, 1989). Woodhead *et al.* (1985) found "anomalously low catches" during periods of high turbidity following a storm at two sites off New York.

Lynch and Johnson (1989) showed that juvenile bluegill in freshwater ponds were significantly less likely to use artificial structures in turbid water with or without predators present. Limitations in the use of artificial structures by visually oriented fishes under high turbidity conditions may partially account for a lack of published studies of fishes on estuarine artificial reefs.

2. Freshwater versus Marine

While principles of ecology are similar in all aquatic systems, freshwater conditions create major differences in artificial reef ecology. Lakes, for example, are closed systems so that recruitment is more likely to be tied to local conditions. There are few instances of reef building organisms in freshwater with the exception of mussels, which create beds or reefs with little vertical relief. Therefore, commonly used structures in freshwater such as brush piles and log cribs do not become enlarged over time, rather they decay and become smaller and less complex. Perhaps because habitat structures tend not to increase in complexity over time, we do not see organisms adapted to make use of the crevices and bore holes that are created during that building process. Thus, the members of a complex marine reef community of algae, arthropods, echinoderms, coelenterates, and fishes is largely missing in freshwater systems. The periphyton community can be relatively diverse taxonomically, however, as illustrated by Rutecki *et al.* (1985) who found 46 taxa of periphyton, most of which were diatoms, on Lake Michigan artificial reefs.

Reduced diversity and niche specificity occurs in freshwater. For example, marine macrofauna in mud show many more phyla and more diversity within phyla than in freshwater (Lopez, 1988); marine zooplankton is more phyletically diverse and species-rich (Lehman, 1988); and macrophyte-grazing fishes are abundant in marine systems and rare in freshwater systems with most of those coming from older tropical lakes and rivers. Also, seagrass-beds have a much higher standing crop than freshwater macrophyte beds (Stevenson, 1988). This reduced diversity in freshwater probably reduces the opportunity to develop complex foraging strategies (e.g., coralivores, pickers, winnowers) noted in marine systems. Finally, the freshwater "reef" community is not well defined spatially, so that participating species and individuals are a functional part of the aquatic community as a whole.

Coral reefs and other "live bottoms" are productive ecosystems in a "sea" of oligotrophic water. In fact, coral reef structure is often dependent on the fact that surrounding water has low nutrient levels since increased nutrients tend to favor algal mats that can smother the coral (see Wiebe, 1988). The fact that these systems provide oases of food and habitat in an immense and relatively infertile oceanic "desert" allows them to support a large variety of reef-adapted life-forms including resident and visiting fishes. Freshwater reefs lack the lush growths of sessile animals that occur on marine reefs. Freshwater rock and brush piles used primarily as fish attractors rarely provide a significantly greater food resource than other parts of the lake.

There simply seem to be fewer organisms available to fill those niches, and there are almost no organisms that are unique to a particular type of natural freshwater habitat structure that would be duplicated by artificial additions. In the Great Lakes of North America, for example, in hundreds of hours of gill netting on artificial reefs in Lake Erie and nonreef reference areas in which over 8000 fish were caught, the only species that was unique to the rock reef was a rock bass (*Ambloplites rupestris*), and it was represented by one individual (Gerber, 1987). Similarly, added rock structures showed few significant differences in fish populations when compared to natural areas of Lake Michigan (Liston *et al.*, 1985) or Lake Ontario (Biener, 1982). Although there were periods of much greater concentrations of fish in the artificially enhanced areas, the variability was so great that statistically significant differences were not observed.

One consistent factor in freshwater seems to be that Centrarchidae are the most common fish family using the structures. Smallmouth bass (*Micropterus dolomieui*), rock bass, bluegill, and white crappie are notorious for their preference for structure, and researchers are able to show significant concentrations of these species at artificial reef areas. Similar studies conducted in a tropical or subtropical marine environment would result in many families and species unique or statistically more abundant in reef areas (e.g., Bohnsack and Talbot, 1980).

The much reduced diversity in freshwater communities and the lack of a specific "reef" community probably contribute to the fact that although nearly 45,000 artificial structures are deployed in U.S. streams, lakes, and reservoirs (McGurrin *et al.*, 1989b), relatively few comprehensive studies have been done on freshwater reefs (reviewed by Brown, 1986). Many factors reported in marine studies to be important are also important in freshwater, such as seasonality, location, habitat complexity, and variation between species (e.g., Prince *et al.*, 1985).

3. Currents

Current conditions are important to the ecology of artificial habitats. Higher coverage and species diversity of marine fouling assemblages are associated with higher current exposure and lower sedimentation (Baynes and Szmant, 1989). Current exposure undoubtedly increases a reef's exposure to larval recruits and may increase the potential food supply. In freshwater, Gannon *et al.* (1985) found exposure to upwelling and fetch influenced reef colonization in Lake Ontario. Nakamura (1985) discussed how pressure waves and current shadows can affect ocean fishes. Pressure waves created by currents impinging on solid reef structures may be sensed by fishes providing them information on the location of a reef out of view.

Fishes can save energy by resting in current shadows. Chang (1985) concluded most fishes were attracted to sites shielded from strong currents. Lindquist and Pietrafesa (1989) showed that fish species abundance was influenced by current vortices.

4. Productivity

Primary productivity may set bounds to the potential total biomass and assemblage structure. This includes not only fouling organisms but also planktonic resources and the surrounding substrate. Randall (1963) reported enhanced fish biomass due to proximity of surrounding seagrass-beds in the Virgin Islands. Scarborough-Bull (1989) reported unusually large size and abundance of mussels (*Mytilus californianus*) on offshore California petroleum platforms due in part to high availability of phytoplankton food resources.

Prince *et al.* (1985) showed that periphyton assemblages attached to the artificial reef structure were considered a major food source for some resident freshwater fishes. In the U.S. Great Lakes, Herdendorf (1985) noted the importance of reefs being shallow enough to be colonized by *Cladophera* and other periphyton species. Unfortunately this leaves little room for artificial reef placement in many freshwater areas. The minimum boat clearance required in open waters of Lake Erie is about 5 m, for example, but in 1985 and 1986 the mean compensation depth (the depth at which oxygen consumption equals oxygen production) was about 7 m. Water quality is considered adequate for most fishes only above the compensation depth.

Whether freshwater artificial habitat structure contributes substantial energy to the fish community is debatable, and certainly depends on latitude and light availability. Prince *et al.* (1985) indicated that sunfishes at artificial reefs showed better body condition and faster growth in weight than fishes caught in other parts of the lake. Pardue (1973) also found growth response to added structures by fish in small containers, but when the experiments were moved to ponds, no response was seen in either bluegill (family Cen-

trarchidae) or tilapia (family Cichlidae) (Pardue and Nielsen, 1979). Johnson
et al. (1988) found no response in fish use of structures in ponds with increasing attached community development, and habitat structure was determined to be used primarily for escape from predators. Also in the United
States, the same conclusion was reached in a study of submerged logs in
Maine (Moring *et al.*, 1989). Similarly, Danehy (1984) found that amphipods
were the major prey item of the yellow perch (*Perca flavescens*) at both
cobble shoals and sand substrates in Mexico Bay, Lake Ontario, and no difference was found between yellow perch stomach contents on artificial reefs
compared to control areas in Lake Michigan (Biener, 1982).

In marine systems, feeding is not necessarily a prime reef attraction for
fishes. Kakimoto (1982) found that little difference existed between the
stomach contents of 10 fish species captured over an artificial reef, a natural
reef, or a controlled fishing ground.

The relationships shown for number of fishes and biomass between marine natural and artificial reefs (Figs. 3.4 and 3.5) imply that the influence
of artificial reef structures is limited by regional productivity. However,
great variability apparently exists within regions. Interesting questions for
artificial reef designers are, What are the upper limits of productivity and,
Can those limits be altered by design features?

IV. Artificial Habitat Design

Part of the challenge to designers of artificial habitat is to raise the productivity envelope set by local limitations to its maximum through clever design
and optimum location. This section discusses design factors that can influence biota. Benthic reefs and FADs are treated separately because of their
many differences. Data come from marine systems, primarily, but data from
freshwater are noted, when available.

A. Benthic Reefs

Part of the reason for the success of benthic artificial reefs in supporting
high densities of organisms is related to increased habitat complexity, defined as the quantity and type of structural elements on a specified spatial
scale (see Bohnsack, 1991). Shulman (1984) showed that increasing habitat
complexity (number of reef units per area) resulted in increased average
numbers of individuals and numbers of species; however, these numbers
leveled off with higher habitat complexity. Gorham and Alevizon (1989) increased habitat complexity with the addition of rope streamers and greatly increased juvenile fish abundance on experimental artificial reefs off of Florida.

Design elements recognized to be most important include physical shape, material composition, surface texture, reef size, and dispersion.

1. Material Composition and Surface Texture

The materials used in artificial reef construction can affect fouling assemblage development. Woodhead and Jacobson (1985) reported some differences in species preferences and rates of colonization between reefs made from concrete blocks and from coal waste materials, although overall fouling assemblages were similar. Fitzhardinge and Bailey-Brock (1989) reported higher coral recruitment to metal surfaces than to rubber in Hawaii, and that concrete provided fouling assemblages most similar to natural coral substrate. Hixon and Brostoff (1985) found that artificial nontoxic polyvinyl-chloride plastic substrate supported the same fouling assemblages as natural dead coral rock substrate for invertebrate abundance and algal biomass, coverage, diversity, and species composition off of Hawaii. Some differences were noted at high grazing intensities due to differences in substrate softness and porosity (surface texture), in which irregular substrate provided refuge from grazing for some settling benthic organisms.

Studies have shown that rough surface texture enhances benthic settlement (Hixon and Brostoff, 1985; Bailey-Brock, 1989) and influences fish composition (Chandler et al., 1985). Hixon and Brostoff (1985) showed that irregular surfaces provided refuge for benthic organisms and supported greater fouling diversity at high fish-grazing intensities. They rejected the hypothesis that irregular surfaces would support greater benthic diversity than flat surfaces at low-grazing intensities because of greater microhabitat partitioning by potentially competing benthic species. Despite these differences, material composition and surface texture seem to be minor factors compared to other variables for most organisms.

2. Shape, Height, and Profile

For this discussion, shape refers to three-dimensional structure; reef height is the distance between the substrate and the highest point on the reef; and profile (or reef relief) refers to how "up-and-down" the reef outline is. All can be important.

Conflicting attitudes exist concerning the importance of the three-dimensional shape of an artificial reef. Some programs, especially those that rely on scrap materials, seem to pay little attention to reef structural shape except in terms of hydrodynamic stability, reef persistence, and potential as a navigational hazard. But as presented in Chapter 4, some programs insist on only carefully engineered and located structures designed for particular species and locations (see also Grove and Sonu, 1985; Grove et al., 1989;

Mottet, 1985; Nakamura, 1985). Practitioners of both approaches seem content with the results based on user satisfaction, although rarely have quantitative biological or economic evaluations been made (McGurrin *et al.*, 1989b; Milon, 1989; Polovina and Sakai, 1989; see also Chapter 7). Japanese programs have been based on a simple comparison of costs of artificial reefs compared to catches (Sato, 1985).

Turner *et al.* (1969) stressed the importance of having 15–18 m diameter open spaces to make artificial reefs attractive to fishes. Walton (1979) found that semienclosed horseshoe-shaped reefs were very successful for attracting flatfishes (order Pleuronectiformes) in Puget Sound, Washington.

In freshwater, specific recommendations for reef shape have been made for U.S. midwestern reservoirs because depth is a major consideration. A row of reef material extending from the 4 m depth contour into deeper water, preferably going down as steep a gradient as available, was recommended to enhance white crappie fishing. A row along the 4 m contour was recommended for bluegill. When both species are desired, managers may find a "T" shape using both patterns useful.

Vertical relief has been correlated with more benthic reef fish species for reefs less than 1 m high (Molles, 1978) but was less effective at elevations greater than 1 m (Patton *et al.*, 1985). On the U.S. Atlantic coast, Stephan and Lindquist (1989) noted little difference in presence of some migratory pelagic species on artificial reefs off North Carolina, despite great differences in the volume of material above 3 m. It is likely that increased exposure to water currents on high profile reefs would favor planktivores over other species. A review of Japanese studies reported that most demersal fishes remain within 3 m of the sea floor and higher reefs might not be effective for these species (Grove and Sonu, 1985). Horizontal extension was more important for attracting demersal fishes, whereas greater vertical extension was more important for midwater species, although reef height did not need to exceed 5 m. Reefs with the same total relief but with nearly vertical sides were better at attracting fishes (Grove and Sonu, 1985).

In freshwater, low profile brush piles and horizontal evergreen trees tended to hold fewer fish than standing evergreens (Lynch and Johnson, 1989). Prince *et al.* (1985) showed that sunfishes favored low profile reef units in shallow water (1.5 m), whereas centrarchid basses preferred high profile units in deeper water (4–6 m) in a Virginia reservoir.

3. Hole Size

For small-holed reefs, little or no effect of hole size or diversity was found on fish species composition or diversity in either the Gulf of California (Molles, 1978) or Australia (Russell *et al.*, 1974; Talbot *et al.*, 1978). Some experimental studies show that hole size and number do affect fish assemblages. Walsh (1985) found that hole composition had little effect on fish assemblages during the day but it was important at night for sheltering fishes off Hawaii. Shulman (1984) found that holes provide shelter from predation and can increase juvenile recruitment, numbers of species, and total fish density on small reefs in the Virgin Islands.

Other studies indicate that reefs with large holes provided less shelter from predation to small fishes, resulting in lower fish abundance and fewer fish species (Shulman, 1984; Hixon and Beets, 1989). Eggleston et al. (1990) found less mortality on small and medium-sized (35 to 55 mm carapace length) spiny lobster, *Panulirus argus*, when heights of artificial shelters (casitas) were reduced from 6 cm to 3.8 and 1.9 cm. The likely mechanism was that predators had less accessibility to lobsters in low shelters. Ogawa (1982) noted that fishes did not inhabit chambers with openings 2 m or larger, and recommended that 0.15 to 1.5 m openings were best for fishery purposes.

Interstice size also has been shown to be important in determining species use in freshwater (e.g., Gannon et al., 1985). Bluegill showed a definite preference for more dense structure; white crappie and largemouth bass seemed to prefer larger openings (Johnson et al., 1988; Lynch and Johnson, 1989).

4. Size

Reef size is an important consideration affecting total volume, bottom coverage, and surface area. As noted previously, small reefs may have insufficient food for certain species to establish permanent populations. However, small reefs may have a higher density than large reefs because they may attract fishes from a proportionally larger area due to a higher perimeter-to-area ratio (Ambrose and Swarbrick, 1989). Many studies have noted higher fish density and biomass at small versus large reefs (e.g., Russell, 1975; Shulman, 1984; Schroeder, 1987; Bohnsack and Sutherland, 1985; DeMartini et al., 1989; Ambrose and Swarbrick, 1989).

In freshwater reservoirs, there tend to be more fishes per square meter (about 40 fish per 8 m²) on small structures. This observation has implications for management because fish can be rather quickly removed by anglers on small structures, while prolonged angling success can be experienced on larger structures. If the intent of managers is to serve as many anglers as possible, the best strategy may be to provide many small structure units; but if the objective is to provide maximum catch rates throughout an angling day, the better approach is larger structures. (The deployment of rows of structure discussed earlier may provide a suitable and more easily marked compromise.)

Depending on the location and species involved, an optimum reef size may exist for fishery purposes. Eggleston *et al.* (1990) found that spiny lobster mortality depends on a relationship between lobster size and shelter size, independent of site; as shelter size increased, more fishes were attracted, which increased mortality on small lobster. Thus, artificial reefs that promote fishes may reduce lobster abundance unless appropriate size and design features are considered. Sato (1985) also discussed optimum reef size. He reported that the minimum effective size for fishery purposes was 400 m^3, and that peak fish harvests occurred with reef bulk volumes of about 3000 m^3/km^2. Harvest declined with larger reefs.

5. Scale

Besides reef size, the spatial and temporal scale used in artificial reef studies is an important and often overlooked consideration. Mapstone and Fowler (1988) noted that controversy and confusion in reef studies can often be traced to differences in the scales in time and space at which researchers operate. Noise at one scale may be critical to processes operating at another scale. For example, different perceptions of assemblage stability have occurred because of different time intervals between samples. Short time intervals between samples implied high variability, whereas longer time intervals implied more stability in artificial reef assemblages (Ogden and Ebersole, 1981; Bohnsack, 1983a).

Similar problems have occurred because of different spatial scales used in research (Sale, 1988; Ambrose and Swarbrick, 1989; Bohnsack, 1991). For example, fish density may be high in small areas around artificial reefs but on larger regional scales effects may be inconsequential. DeMartini *et al.* (1989) noted how artificial reef population estimates based on density can be misleading. Although the density of fishes off California was much higher on an artificial reef than on natural reefs, the total standing stock was 60 times greater on natural reefs simply because of the larger bottom area covered. In this case, artificial reefs appeared to have less importance on a regional scale than when examined on a scale restricted to the immediate area around an artificial reef site. Ambrose and Swarbrick (1989) came to similar conclusions in a broader study off California.

Petroleum extraction structures, in contrast, potentially could greatly increase the abundance of hard bottom fauna in the northern Gulf of Mexico because of their abundance, distribution, and size. Over 4000 structures exist (Ditton and Auyong, 1984) and a single structure in 30 m of water can have 0.81 hectare of hard surface exposed to the water column (Shinn, 1974). A structure in 45 m can have 1.2–1.6 hectares of surface (Bull, 1990). These platforms comprise a significant amount of available hard surface, estimated as 15 to 28% (Bull, 1990) of the limited, naturally available, hard substrate (Parker *et al.*, 1983; Rezak *et al.*, 1985).

6. Dispersion

The arrangement of reef materials within or between reefs can be important ecologically. In Japan, large nonreef areas are left between artificial reefs to maximize fish production (Nakamura, 1985). In the United States, California rock reefs are frequently constructed in multiple modules (see Ambrose *et al.*, 1989). Because many fishes feed away from a reef or on passing plankton, reef materials that are too concentrated may limit plankton availability or may lead to overgrazing for surrounding bottom. Roving predators may move more frequently between reefs that are closer together, which may be an advantage or disadvantage depending on management goals (Bohnsack, 1989). Possibly, ecotonal species are favored by dispersed patterns, whereas reef-feeding species are favored by clumped patterns, according to our observations.

Turner *et al.* (1969) noted that to sustain sportfishing, large open spaces should be incorporated into artificial reef design. Hueckel *et al.* (1989) reported success by arranging 14 artificial reef structures to form one reef so as to maintain a 1:2 ratio of reef material to sand bottom in Puget Sound, Washington, and in order to maintain trophic relationships between reef fauna and surrounding bottom. Brock and Norris (1989) compared four reef designs off Hawaii and concluded that haphazardly deployed materials provided significantly poorer enhancement of fish assemblages than reefs constructed of designed modules assembled into a specific configuration. However, reef dispersion was confounded with other variables including reef height, materials, and age. Brock and Norris (1989) suggested that an inverse relationship exists between reef material dispersion and the mean size of resident fishes.

Despite these suggestions regarding the importance of dispersion patterns on artificial reef communities, no conclusive experimental studies exist showing that particular patterns are best.

B. FADs

FADs are objects suspended in the water column or floated at the surface to attract fishes, as illustrated in Chapter 1. They can be used alone or suspended above a benthic reef. FADs exploit a natural tendency for pelagic fishes to orient toward structure. Some fishes may use them for orientation to currents, and some predators may use them as spatial reference points for pelagic searches. In oceanic environments, FADs may be attractive to small fishes because they mimic naturally floating objects that get trapped in oceanographic convergence zones where food also is concentrated.

The ecology of FADs may differ somewhat from most bottom reefs in that fish assemblages are probably more ephemeral. They are unlikely to directly provide much food, although predators may feed on small fishes attracted to them. Rountree (1989) reviewed evidence supporting various hypotheses on why fishes associate with FADs, and in an experimental study reported a linear effect of fish abundance with FAD size for some species in the Atlantic off South Carolina. He concluded that prey fishes associate with drifting objects to escape predation. Shadows provided by FADs may confuse some predators and provide camouflage to deter attacks from below. Helfman (1979) noted how shadows from floating rafts aid the visual ability of lake fishes within the shadow. Beets (1989) reported that FADs suspended above bottom reefs at 26 m depth off the Virgin Islands increased larval recruitment to the bottom reefs. An early experimental FAD is depicted in Fig. 3.6.

Catch rates around FADs are frequently higher than in control areas without FADs. Buckley *et al.* (1989) reported significantly higher catch rates at FADs off American Samoa than from open water areas, but no significant difference between FADs and offshore bank areas. Feigenbaum *et al.* (1989) reported higher catches at FADs than from control areas off Puerto Rico but

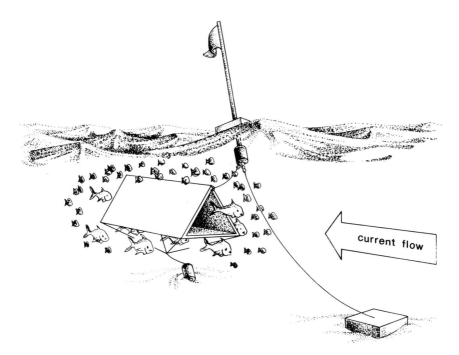

Figure 3.6 An experimental design of a FAD. Research on FADs in recent decades is supplanted by centuries of use in fisheries settings (after Klima and Wickham, 1971).

concluded that FADs were unlikely to dramatically increase harvest. However, Smith *et al.* (1981) examined FADs suspended in freshwater lakes in Alabama and found low numbers of harvestable fish in all areas, although a greater number of fishes utilized structures off points of land than in coves.

Despite the success of FADs for improving harvest, an unresolved ecological question, important to fishery managers, is how fishing at FADs affects specific fish populations. If FADs reduce natural mortality rates by providing protection from predators, then they may be substituting fishing mortality for natural mortality. Effects probably depend on the species involved. If FADs only aggregate fish and make them easier to catch, they could more easily lead to overexploitation. Samples and Sproul (1985) modeled fishing at FADs and conclude that without regulations FADs are unlikely to "generally increase fishermen's aggregate profit position (p. 305)." Potential problems were especially acute when fishing costs were low, fishing was efficient, and FADs were effective at aggregating fishes. Interestingly, the same arguments could apply to benthic reefs, especially when they act primarily as fish aggregators. Discussion of fishery impacts is presented in Chapter 5.

V. Population Dynamics

A. Colonization and Recruitment Patterns

Colonization refers here to the establishment of populations on artificial reefs; recruitment refers to the resupply of established populations. Colonization usually begins at the moment of deployment. Most fouling organisms colonize artificial reefs by settlement of pelagic larvae. Fishes can colonize either by direct settlement of pelagic larvae or by immigration of juveniles or adults. Many reef fishes first utilize some intermediate habitat before using reef habitats as older juveniles or adults. Russell *et al.* (1974), working with tropical reefs, found most initial colonization was by juveniles. Gascon and Miller (1981), working in temperate areas, found most fish colonization was by adult and subadult individuals. The latter observation, if it proves to be generally applicable, may be an important difference in recruitment ecology between tropical and temperate artificial reefs.

Colonization by individual species can occur rapidly or slowly. The time of reef deployment may influence initial colonization depending on what recruits are available in the water column. Ceramic plates hung under docks near Beaufort, North Carolina, for example, developed different fouling assemblages depending on the month of exposure (Sutherland, 1974; Sutherland and Carlson, 1977). Russell *et al.* (1974) and Talbot *et al.* (1978) showed that fishes colonized artificial reefs off Australia more slowly in winter than in summer. Among fouling organisms, species that first occupy available attachment sites may effectively prevent later settlement by other species. Resident fishes may influence recruitment by predation on potential recruits (Shulman *et al.*, 1983; Bohnsack, 1991). Shulman *et al.* (1983) also showed that larval settling by three tropical reef fish species significantly decreased in the presence of territorial damselfish.

Immigration to artificial reefs often occurs by pelagic phases early in the life cycle of many species, especially those characterized by sessile adulthood (Schoener, 1982). Most marine reef organisms have a bipartite life cycle with a planktonic egg or larval dispersal phase prior to recruiting to a reef. Marine fishes, for example, may spend from a few days to over 100 days in pelagic stages (Doherty and Williams, 1988; Mapstone and Fowler, 1988). Osman (1978) concluded that about 90% of colonists in the marine realm must immigrate as larvae. Mortality is great during this pelagic dispersal phase because of exposure to predation and environmental vicissitudes in which larvae can starve or be carried away from suitable settlement habitat by water currents. In the marine environment, offspring probably rarely settle on the same reef from which they were spawned because of pelagic dispersal (Doherty and Williams, 1988). In freshwater, however, there is a greater chance for direct recolonization of parent habitats, especially in smaller bodies of water.

Substrates are colonized differently between localities but also within a locality even when identical substrate is used (Schoener, 1982). There are several reasons for this phenomenon. The availability of settlers may vary temporally depending on when a reef is made available for colonization. Subtle differences in placement may affect settlement and initial colonization due to chance events, clumping of settlers, or microenvironmental gradients. Also, differences in initial colonization can lead to major differences in resulting assemblages because of different abundances and types of settling predators. Extensive studies conducted on successional processes in natural habitats are beyond the scope of this chapter but may provide insight into successional processes on artificial structures (e.g., Vance, 1988; Hairston, 1989).

Recruitment primarily comes from external sources using the same mechanisms as colonization with the exception that some solitary and colonial sessile organisms (e.g., sponges, anemones, algae) can spread vegetatively after settlement. Successful recruitment and population persistence depends on the presence of suitable conditions and on dynamic factors discussed later. The supply of recruits can be continuous or discontinuous.

Often recruitment can vary unpredictably in space and over time due to changes in population reproductive output and environmental uncertainties (Doherty and Williams, 1988; Anderson et al., 1989).

1. Successional Processes

Artificial reef colonization is often very rapid and shows similar and repeatable patterns (Bohnsack and Talbot, 1980; Schoener, 1982; Bohnsack and Sutherland, 1985). Numerous descriptive accounts document colonization patterns and variation for specific areas (e.g., Turner et al., 1969; Russell, 1975; Stone et al., 1979). These colonization patterns can be described quantitatively even though differences in species composition are discernable. Common similarities observed between sites in an area are speed of colonization and percentage of cover (Schoener, 1982).

Assemblage structure may be affected by competition, predation, and by physical disturbance (Schoener, 1982). After initial colonization, populations often fluctuate cyclically or seasonally around some average value (Figs. 3.7 and 3.8). In southern Florida, for example, reefs of small concrete cubes (132 cm per side) were rapidly colonized. Small reefs averaged more individuals per module than larger reefs (Fig. 3.7). Abundance of fishes fluctuated greatly with seasons, although fluctuations were dampened on larger reefs. Variation was high within replicates. The patterns of numbers of individuals may be compared with changes in biomass (Fig. 3.8). Rapid biomass accumulation shows the importance of aggregation during the initial colonization phase. Biomass was proportionally less variable on larger reefs and showed less seasonal fluctuation than numbers of individuals.

Succession is a controversial topic for many communities (Hairston, 1989). Succession can be defined as a descriptive account of assemblage changes over time. In the strict sense, succession is defined as "replacement of populations in a habitat through a regular progression to a stable state" (Ricklefs, 1973, p. 794). Disagreement exists as to the importance and existence of a stable state or equilibrium in most marine communities (Chesson and Case, 1986). Some assemblages may be in nonequilibrium because a stable state is never achieved due to periodic physical or biological disturbance (e.g., Talbot et al., 1978). Gascon and Miller (1981) found a lack of competitive interactions among fishes on a temperate reef as evidence supporting a nonequilibrium model. Wendt et al. (1989) examined fouling assemblages on five artificial reefs of different age off of South Carolina and found no consistent patterns with age for reefs between 3.5 and 10 years of age. Carter et al. (1985) found successional differences among sessile biota between artificial reefs and natural reefs off southern California, and they suggested that succession may be affected by the time of reef placement and isolation from other natural reefs in the area.

CONTROL SITES
(0 MODULES)

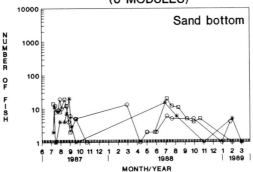

TWO MODULE SITES

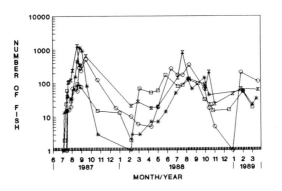

EIGHT MODULE SITES

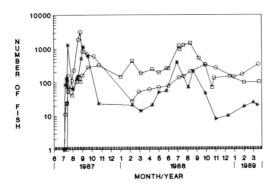

Figure 3.7 Fish colonization and seasonal variation in numbers of individuals as a function of reef size at artificial reefs deployed off Miami, Florida. Sand bottom and large natural reef control sites show numbers observed within a 7.5 m circular radius from a stationary diver (Bohnsack and Bannerot, 1986). Vertical bars show 95% confidence intervals. The smaller control patch reef covered approximately 28 m². (*Figure continues.*)

ONE MODULE SITES

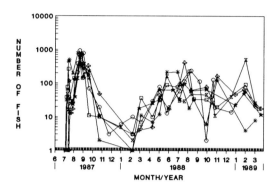

FOUR MODULE SITES

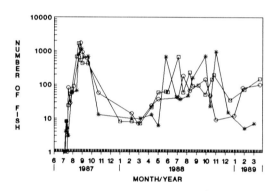

NATURAL REEF SITES

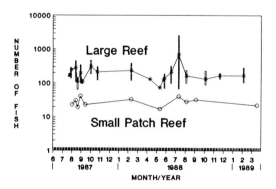

Figure 3.7 (*Continued*)

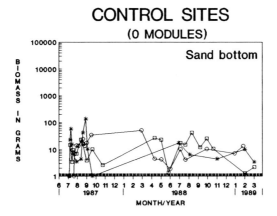

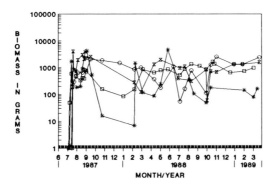

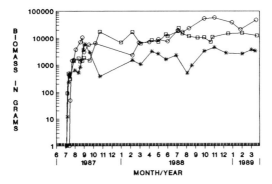

Figure 3.8 Colonization and seasonal variation in fish biomass, as a function of reef size, at artificial reefs deployed off Miami, Florida. (*Figure continues.*)

ONE MODULE SITES

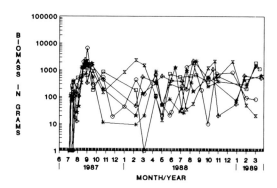

FOUR MODULE SITES

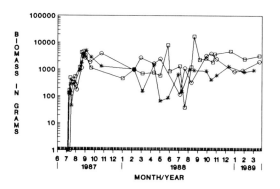

NATURAL REEF SITES

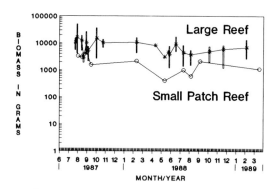

Figure 3.8 (*Continued*)

2. Island Biogeographic Theory

Island biogeographic theory (IBT) is presented as an example of how ecological theory can be applied to artificial habitats. IBT has frequently been used to model the colonization and the dynamics of artificial reef assemblages for various sessile organisms (MacArthur and Wilson, 1967; Schoener, 1982) and fishes (e.g., Nolan, 1975; Molles, 1978; Bohnsack, 1979). According to IBT, the number of species in an assemblage at an artificial reef is in dynamic equilibrium between colonization and extinction (Fig. 3.9). Initially, new artificial reefs have the highest colonization rates as species colonize an unoccupied habitat; there is no extinction because there is nothing to go extinct. Over time, the colonization rate decreases because fewer species are available to colonize an already occupied reef, and successful colonization becomes more difficult due to competition and predation from residents. As species colonize, the extinction rate should rise. A reef with all possible species present could have no colonization because there are no species left to colonize; extinction rates should be at their highest levels from intense competition and predation among residents. Due to stochastic (chance) events, small populations should have higher probabilities of becoming extinct than large populations (MacArthur and Wilson, 1967).

Island biogeographic theory predicts an equilibrium process, although the assemblages do not have to be in static equilibrium. In fact, the theory predicts a dynamic equilibrium with high flux and turnover, where turnover

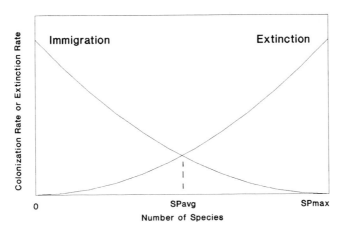

Figure 3.9 The basic island biogeographic model. The average number of species SPavg on a reef is determined by a balance between immigration and extinction rates. SPmax is the total number of species potentially available to colonize a reef. (Modified from MacArthur and Wilson, 1967).

is the total number of colonizations and extinctions over a period of time (MacArthur and Wilson, 1967). Any exogenous changes in colonization or extinction rates, such as seasonal recruitment or changes in predation, would affect the assemblage.

The advantage of ecological models such as IBT is that generalizations and predictions can be made about the dynamics of artificial reef assemblages that may be unexpected or not otherwise possible. For example, initial colonization rates should predict turnover rates; large reefs should have more species and lower turnover than small reefs; and reefs remote from recruitment sources should have fewer species than reefs of the same size closer to recruitment sources. Large reefs should have more species because of larger population sizes and lower extinction rates. Reefs close to recruitment sources should have higher colonization rates and thus more species than a distant reef of the same size. These simple relationships can be combined into more complex models (Fig. 3.10). The IBT model could be refuted by demonstrating that assemblages are static over time, but this does not seem to be the case with most reefs studied (although see Lukens, 1981). An alternative to IBT is to treat each reef as a unique entity with no similarity to other reefs, although this approach seems less satisfying.

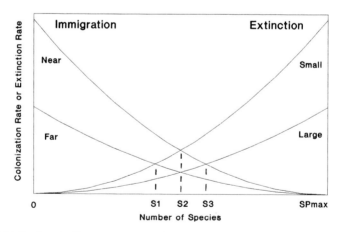

Figure 3.10 Predicted effects on mean number of species of artificial reef size and distance from colonization sources according to Island Biogeographic Theory. (Note that small, distant reefs should have the fewest species (S1), while large, near reefs should have the most species (S3). Small, near reefs and large, distant reefs should have an intermediate number of species (S2). SPmax is the total number of species potentially available to colonize a reef. (Modified from MacArthur and Wilson, 1967).

B. Population Regulation

The composition of any artificial reef assemblage can be viewed as the result of ecological forces that regulate population abundance. These forces are often treated as gradients of opposing forces. Major recognized gradients include the relative importance of physical versus biological limitation; density-dependent versus density-independent mortality; deterministic versus stochastic effects; and habitat versus recruitment limitation.

1. Physical versus Biological Regulation

Classically, temperate environments are thought be more regulated by physical (abiotic) processes, whereas tropical environments are regulated by biological (biotic) processes (Menge and Sutherland, 1976, 1987). More extreme seasonality at higher latitudes is an example of this gradient. Among abiotic forces, normal or extreme physical conditions can regulate population size. Depth, excessive turbidity, or a lack of nutrients, for example, can limit algal growth. Conversely, water mixed by storms may provide nutrients that promote algal growth. Lack of water movement can limit populations of sessile filter feeding organisms within the fouling assemblage, whereas prevailing strong currents can promote their growth. Ambrose and Anderson (1990) noted that abundances of some species in soft-bottom sediments in California may change near artificial reefs because of sediment grain size changes due to physical processes interacting with the presence of an artificial reef. Severe physical disturbances such as storms, low oxygen levels, and extreme cold or hot spells can periodically kill certain organisms and limit population size. Physical factors are frequently (but not always) considered density-independent factors because their impact may be independent of population size.

When abiotic environmental conditions are stable, biological factors are more likely to be important for limiting population sizes and assemblage composition (Menge and Sutherland, 1976). The two most important biological processes are competition and predation, where predation includes grazing, disease, and parasitism. Biotic factors are frequently considered density-dependent processes because their effect changes with population size. For example, as a population increases, competition should become more important. Likewise, the effect of predation on prey abundance will increase as predator abundance grows. Moderate disturbance, as opposed to extreme disturbance, may increase the number of species present by reducing populations of dominant predators or competitors and allowing less competitive or less vulnerable species to colonize and persist (Connell, 1978).

Competition and predation can influence the abundance of populations in complex ways (Werner, 1986). Fish grazing can affect fouling assemblages

(Hixon and Brostoff, 1985; Fizthardinge and Bailey-Brock, 1989) (Fig. 3.11). Davis *et al.* (1982) concluded that sea pens (*Stylatula undulata*) were eliminated around artificial reefs, probably due to grazing, while populations of tube-building polychaetes were enhanced (but see Ambrose and Anderson, 1990). Alevizon and Gorham (1989), however, did not detect changes in surrounding nonreef fish communities following the addition of an artificial reef in the Florida Keys. Ambrose and Anderson (1990) concluded that the overall effect of a California artificial reef on the surrounding infauna was limited to a small area near the modules.

Fouling studies have shown that biologically generated habitat complexity may explain some differences in species composition between sites. For example, structural components of organisms can provide canopy and protection for others. Russ (1980) showed that arborescent (tree-like) forms of bryozoans reduced the foraging efficiency of fish and reduced the importance of predation in structuring the fouling assemblage. Choat and Ayling (1987) found differences in fish composition on natural reefs in New Zealand were related to differences in the availability of food caused by the presence or absence of laminarian and fucoid algae. Holbrook *et al.* (1990) showed differences in fish assemblages off California were related to the presence or absence of giant kelp, but otherwise only weak differences in species composition existed among reefs of different habitat types.

2. Density-Dependent and Density-Independent Mortality

Density-independent mortality generally destabilizes populations because its impact is independent of population size. For example, a severe cold spell may kill 80% of a population at an artificial reef regardless of whether the population is large or small. Density-dependent mortality is usually stabilizing in that it changes with population and it forces abundance to remain within a range of possible levels. For example, assuming planktonic food availability is relatively constant, small populations of sessile planktivores may grow because excessive food is available. When the population gets large, food per individual may be scarce; increased competition between individuals results in greater direct mortality, greater risks of predation, or perhaps increased individual vulnerability to disease. Thus, at high densities populations may slow, stop growing, or even decline. Other examples of both density-dependent and density-independent forces are presented in the previous section, and in Chapter 6.

How artificial reefs interact with these forms of mortality can be an important consideration. Structural components that provide shelter from predation may limit the density-dependent effects of predation but may do nothing to protect organisms from extreme temperature-induced mortality.

Figure 3.11 Differences between fouling assemblages due to fish grazing. The top reef, composed of two modules, had few resident fishes and thicker growths of algae compared to the bottom reef, composed of eight modules, which had many resident grazers. Note the differences in algae along the top of modules. Both reefs were deployed at the same time and photographed approximately 2.5 yr after deployment. (Photo credit, top: courtesy of J. A. Bohnsack; bottom: courtesy of M. W. Hulsbeck.)

Artificial reefs that attract unusually high densities of fishes may lead to increased natural predation or attract higher fishing effort, limiting its effectiveness. Certainly, knowledge of the habitat requirements and factors that limit the population size of targeted species is important for optimum artificial habitat design and placement.

3. Deterministic versus Stochastic Factors

Considerable research has treated the relative importance of deterministic versus stochastic factors in tropical coral reef fishes (the order versus chaos controversy) (e.g., Smith, 1977; 1978; Sale, 1977; Sale and Williams, 1982). The deterministic school views reef assemblages as evolutionarily, coevolved communities with high predictability, stability, and a biologically accommodated community structure. The opposing stochastic view emphasizes the variability and unpredictability of assemblages caused by chance factors. Some elements in both of these viewpoints are probably true and have relevance for artificial reefs (Bohnsack, 1983a). Deterministic elements imply that specific design features can be used to benefit specific targeted species. Otherwise, assemblages may be formed due to chance, with little human influence. Some design features appear to specifically benefit, for example, abalone and lobster (Grove and Sonu, 1985). Except for features discussed earlier, such as hole size, which appear to benefit broad categories of predators or prey, most claims that specific structures benefit certain fish species have not been experimentally verified.

4. Habitat versus Recruitment Limitation

In recent years, the order versus chaos controversy seems to have been recast in terms of asking under what conditions populations are limited by habitat availability (an ordering factor) or recruitment supply (presumably a stochastic factor). Resolution of this question is essential for understanding the ecology of artificial and natural reefs, and is applicable to invertebrates (e.g., Underwood and Fairweather, 1989) and fishes (e.g., Doherty and Williams, 1988). This habitat versus recruitment controversy may be the most important issue in artificial habitat application.

Under habitat limitation, increasing the amount of habitat can directly increase population abundance. When habitat is not limiting, populations can be restricted by the available supply of larvae, postsettlement mortality (having nothing to do with habitat availability), or fishing mortality (Shulman and Ogden, 1987; Richards and Lindeman, 1987; Doherty and Williams, 1988; Polovina and Sakai, 1989). Note that recruitment supply can be limited by either biotic or abiotic events. Thus for habitat-limited species, deployment of artificial reefs can actually increase the total amount of reef fishes and not just make them more available by changing their distribution. For recruitment-limited species, artificial reefs may have little effect on increasing total abundance and under some conditions can reduce populations by making fishes easier to catch (Bohnsack, 1989).

Most scientific thinking has shifted over the last 20 years from emphasis on habitat as the most important factor limiting population abundance to emphasis on the importance of recruitment variability (Richards and Lindeman, 1987; Doherty and Williams, 1988; Warner and Hughes, 1988). Bohnsack (1989) produced a model predicting that artificial reefs are most likely to increase total biomass or numbers of fish for reef species that are habitat limited, demersal, philopatric, territorial, and obligatory relative to species that are recruitment limited, pelagic, highly mobile, partially reef-dependent or opportunistic (Fig. 3.12). Although some of these predictions

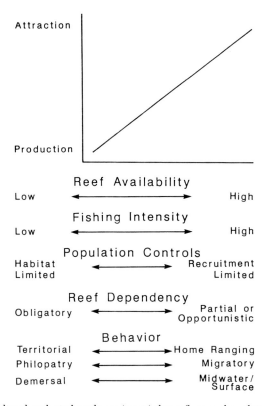

Figure 3.12 Predicted ecological gradients (x-axis) that influence the relative contributions of fishes attracted or produced by an artificial habitat. Conditions at the left favor new biomass production, whereas conditions on the right favor attraction of existing fishes. A linear response is shown only for illustrative purposes. (From Bohnsack, 1989).

have been tested (e.g., Polovina and Sakai, 1989), there is a need for more extensive quantitative testing in different areas and perhaps refinement or rejection of parts of the model. As noted elsewhere, many of the experimental studies have been of tropical coral reef fauna, with the possibility that this process may differ for temperate or freshwater reef fauna. Little information exists on the relative importance of recruitment or habitat limitation on freshwater artificial reefs.

5. Fishing Mortality

Fishing mortality can be an important influence on artificial reef assemblages because artificial habitats are often constructed for fishing purposes and frequently have higher fishing mortality than natural reefs (Bohnsack and Sutherland, 1985; Buckley and Hueckel, 1985; Matthews, 1985; Solonsky, 1985). Polovina and Sakai (1989) warned that managers need to consider that the benefits of artificial reefs may vary with the level of fishing effort. If fishing mortality is too high, the benefits of artificial habitats can be dissipated or overfishing problems can be aggravated (Bohnsack, 1989). Little research has been directed at this problem (see Chapter 5).

VI. Conclusion

This chapter has provided an overview of factors that influence artificial habitat ecology. Clearly, artificial reef assemblages are dynamic and respond to the same forces that act on natural reefs. In recent years, great strides have been made in increasing our understanding of artificial habitat ecology, although many questions remain unanswered. Obviously many controversies and uncertainties exist concerning successional processes and the relative importance of recruitment and habitat limitation, deterministic and stochastic events, competition and predation, and whether assemblages act as equilibrium or nonequilibrium systems.

Part of our lack of understanding of artificial reef ecology is that we do not fully understand the ecology of natural reefs. Interestingly, much of what we know about natural reef ecology is based on experimental studies using artificial habitats, which can be more easily manipulated than natural habitats.

Some of the problems impeding progress include the variety of locations and materials used in artificial reefs, the diversity of artificial reef assemblages, and the complexity of biological and physical interactions. Clearly, there is a need for more controlled experimental studies that test ecological theory, reef function, and design performance. Also helpful would be comparative studies that further elucidate ecological differences between artificial reefs in temperate and tropical settings or in freshwater or marine environments.

The limitations of sampling methodology remain a major obstacle to our understanding of artificial reef ecology. A need clearly exists for inexpensive and more reliable sampling techniques (see Chapter 6). The tendency to rely only on numbers of individuals as a measure of "ecological currency" should be questioned because this greatly underplays other important parameters of ecological importance such as size, function, and frequency-of-occurrence.

Finally, management goals should be more precisely stated in an ecological context in order to direct ecological evaluations. Vague objectives such as "improving the marine environment" are untestable (and unacceptable) compared to objectives of "increasing the standing stock of X, Y, and Z species by $N\%$" in a given area.

A key benefit of artificial habitats is our ability to more easily manipulate assemblages and perhaps improve the habitat with appropriate artificial habitat use. In general, location appears to be more important at influencing assemblages than reef structure, although this should not imply that optimization of designed structure would not benefit fisheries or the habitat. Apparently, regional limits to the biological productivity of artificial habitats exist, although great variability exists within a region. It remains to be seen how much design and placement of artificial reefs can exploit this variability.

The most important problems remaining to be addressed deal with the ecological implications of the best use of artificial habitats and their proper role in fisheries management and habitat mitigation. Because artificial reef construction is an active management step, it is crucial that we understand the implications of that decision. Understanding more about artificial reef ecology can lead to significant improvements in applying artificial habitats to resource management problems: improved knowledge can (1) provide convincing evidence about when it is appropriate, or inappropriate, to use artificial habitats; (2) serve to quantify the magnitude of the benefits that accrue from an artificial habitat; and (3) improve the design and placement of artificial habitats to maximize benefits. The study of artificial habitat ecology is still a very young science and it promises great advances in our understanding and use of the aquatic environment.

References

Alevizon, W. S., and J. C. Gorham. 1989. Effects of artificial reef deployment on nearby resident fishes. Bulletin of Marine Science 44:646–661.

Alevizon, W. S., J. C. Gorham, R. Richardson, and S. A. McCarthy. 1985. Use of man-made reefs to concentrate snapper (Lutjanidae) and grunts (Haemulidae) in Bahamian waters. Bulletin of Marine Science 37:3–10.

Ambrose, R. F., and T. W. Anderson. 1990. The influence of an artificial reef on the surrounding infaunal community. Marine Biology 107:41–52.

Ambrose, R. F., and S. L. Swarbrick. 1989. Comparison of fish assemblages on artificial and natural reefs off the coast of southern California. Bulletin of Marine Science 44: 718–733.

Ambrose, R. F., D. C. Reed, J. M. Engle, and M. F. Caswell. 1989. California comprehensive offshore resource study: Summary of biological resources. Report to the California State Lands Commission.

Anderson, T. W., E. E. DeMartini, and D. A. Roberts. 1989. The relationship between habitat structure, body size and distribution of fishes at a temperate artificial reef. Bulletin of Marine Science 44:681–697.

Bailey-Brock, J. H. 1989. Fouling community development on an artificial reef in Hawaiian waters. Bulletin of Marine Science 44:580–591.

Baynes, T. W., and A. M. Szmant. 1989. Effect of current on the sessile benthic community structure of an artificial reef. Bulletin of Marine Science 44:545–566.

Beets, J. 1989. Experimental evaluations of fish recruitment to combinations of fish aggregating devices and benthic artificial reefs. Bulletin of Marine Science 44:973–983.

Biener, E.W. 1982. Evaluation of an artificial reef placed in southeastern Lake Michigan: Fish colonization. Master's Thesis, Michigan State University, East Lansing.

Bohnsack, J. A. 1979. The ecology of reef fishes on isolated coral heads: An experimental approach with emphasis on island biogeographic theory. Ph.D. Dissertation, University of Miami, Miami, Florida.

Bohnsack, J. A. 1983a. Species turnover and the order versus chaos controversy concerning reef fish community structure. Coral Reefs 1:223–228.

Bohnsack, J. A. 1983b. Resiliency of reef fish communities in the Florida Keys following a January 1977 hypothermal fish kill. Environmental Biology of Fishes 9:41–53.

Bohnsack, J. A. 1989. Are high densities of fishes at artificial reefs the result of habitat limitation or behavioral preference? Bulletin of Marine Science 44:632–645.

Bohnsack, J. A. 1991. Habitat structure and the design of artificial reefs. Pp. 412–426 in S.S. Bell, E. D. McCoy, and H. R. Mushinsky, (eds.). Habitat structure: The physical arrangement of objects in space. Chapman & Hall, London.

Bohnsack, J. A., and S. P. Bannerot. 1986. A stationary visual census technique for quantitatively assessing community structure of coral reef fishes, NOAA Technical Report NMFS 41. U.S. Department of Commerce, Miami, Florida.

Bohnsack, J. A., and D. L. Sutherland. 1985. Artificial reef research: A review with recommendations for future priorities. Bulletin of Marine Science 37:11–39.

Bohnsack, J. A., and F. H. Talbot. 1980. Species-packing by reef fishes on Australian and Caribbean reefs: An experimental approach. Bulletin of Marine Science 30:710–723.

Bohnsack, J. A., D. E. Harper, D. B. McClellan, and M. Hulsbeck. 1989. The relative importance of recruitment, attraction, and production of reef fishes on natural and modular artificial reefs, Project Report. Florida Sea Grant, Gainesville.

Brock, R. E., and R. A. Grace. 1987. Fishery enhancement through artificial reef development for nearshore Hawaiian waters, Final report, Cooperative Agreement NA-85-ABH-00028. Hawaii Institute of Marine Biology, University of Hawaii, Honolulu.

Brock, R. E., and J. E. Norris. 1989. An analysis of the efficacy of four artificial reef designs in tropical waters. Bulletin of Marine Science 44:934–941.

Brown, A. M. 1986. Modifying reservoir fish habitat with artificial structures. Pp. 98–102 in G. E. Hall and M. J. Van Den Avyle, editors. Reservoir fisheries management: Strategies for the 80's. Reservoir Committee, Southern Division, American Fisheries Society, Bethesda, Maryland.

Buckley, R. M. 1989. Habitat alterations as a basis for enhancing marine fisheries. California Cooperative Fishery Investigations, CalCOFI Report 30:40–45.

Buckley, R. M., and G. J. Hueckel. 1985. Biological processes and ecological development on an artificial reef in Puget Sound, Washington. Bulletin of Marine Science 37:50–69.

Buckley, R., J. Grant, and J. Stephens, Jr. 1985. Third International Artificial Reef Conference: 3–5 November 1983, Newport Beach, California. Bulletin of Marine Science 37:1–2.

Buckley, R. M., D. G. Itano, and T. W. Buckley. 1989. Fish aggregation device (FAD) enhancement of offshore fisheries in American Samoa. Bulletin of Marine Science 44:942–949.

Bull, A. S. 1990. Fish assemblages at oil and gas platforms, compared to natural hard/live bottom areas in the Gulf of Mexico. Coastal Zone '90:979–987.

Burress, R. M. 1961. Fishing pressure and success in areas of flooded standing timber in Bull Shoals Reservoir, Missouri. Proceedings of the Annual Conference of the Southeastern Association of Game and Fish Commissioners 15:296–298.

Cairns, J., Jr., editor. 1982. Artificial substrates. Ann Arbor Science Publishers, Ann Arbor, Michigan.

Carter, J. W., A. L. Carpenter, M. S. Foster, and W. N. Jessee. 1985. Benthic succession on an artificial reef designed to support a kelp-reef community. Bulletin of Marine Science 37:86–113.

Chandler, C. R., R. M. Sanders, Jr., and A. M. Landry, Jr. 1985. Effects of three substrate variables on two artificial reef fish communities. Bulletin of Marine Science 37:129–142.

Chang, K. 1985. Review of artificial reefs in Taiwan: Emphasizing site selection and effectiveness. Bulletin of Marine Science 37:143–150.

Chang, K., S. Lee, and K. Shao. 1977. Evaluation of artificial reef efficiency based on the studies of model reef fish community installed in northern Taiwan. Bulletin of the Institute of Zoology, Academia Sinica 16:23–36.

Chesson, P. L., and T. J. Case. 1986. Overview: Nonequilibrium community theories: Chance, variability, history, and coexistence. Pp. 229–239 in J. Diamond and T. J. Case, editors. Community ecology. Harper & Row, New York.

Choat, J. H., and A. M. Ayling. 1987. The relationship between habitat structure and fish faunas on New Zealand reefs. Journal of Experimental Marine Biology and Ecology 110:257–284.

Connell, J. H. 1978. Diversity in tropical rain forests and coral reefs. Science 199:1302-1310.

Crumpton, J. E., and R. L. Wilbur. 1974. Florida's fish attractor program. Pp. 39–46 in L. Colunga and R. B. Stone, editors. Proceedings of an International Conference on Artificial Reefs, TAMU-SG-74-103. Texas A&M University Sea Grant College Program, College Station, Texas.

Danehy, R. J. 1984. Comparative ecology of fishes associated with natural cobble shoals and sand substrates in Mexico Bay, Lake Ontario. Master's Thesis, State University of New York, Syracuse.

Davis, N., G. R. Van Blaricom, and P. K. Dayton. 1982. Man-made structures on marine sediments: Effects on adjacent benthic communities. Marine Biology 70:295–303.

DeMartini, E. E., D. A. Roberts, and T. W. Anderson. 1989. Contrasting patterns of fish density and abundance at an artificial rock reef and a cobble-bottom kelp forest. Bulletin of Marine Science 44:881–892.

D'Itri, F., editor. 1985. Artificial reefs: Marine and freshwater applications. Lewis Publishers, Inc., Chelsea, Michigan.

Ditton, R. B., and J. Auyong. 1984. Fishing offshore platforms—central Gulf of Mexico: An analysis of recreational and commercial fishing use at 164 major offshore petroleum structures, OCS Monograph MMS 84-0006. U.S. Department of the Interior, Minerals Management Service. Metairie, Louisiana.

Doherty, P. J., and D. M. Williams. 1988. The replenishment of coral reef fish populations. Annual Review of Oceanography and Marine Biology 26:487–551.

Eggleston, D. B., R. N. Lipcius, D. L. Miller, and L. Coba-Cetina. 1990. Shelter scaling regulates survival of juvenile Caribbean spiny lobster *Panulirus argus*. Marine Ecology Progress Series 63:79–88.

Fast, D. E., and F. A. Pagan. 1974. Comparative observations of an artificial tire reef and natural patch reefs off southwestern Puerto Rico. Pp. 49–50 *in* L. Colunda and R. Stone, editors. Proceedings: Artificial reef conference. Texas A&M University, TAMU-SG-74-103.

Feigenbaum, D., A. Friedlander, and M. Bushing. 1989. Determination of the feasibility of fish attracting devices for enhancing fisheries in Puerto Rico. Bulletin of Marine Science 44:950–959.

Fitzhardinge, R. C., and J. H. Bailey-Brock. 1989. Colonization of artificial reef materials by corals and other sessile organisms. Bulletin of Marine Science 44:567–579.

Gannon, J. E., R. J. Danehy, J. W. Anderson, G. Merritt, and A. P. Bader. 1985. The ecology of natural shoals in Lake Ontario and their importance to artificial reef development. Pp. 113–140 *in* F. D'Itri, editor. Artificial reefs: Marine and freshwater applications. Lewis Publishers, Inc., Chelsea, Michigan.

Gascon, D., and R. A. Miller. 1981. Colonization by nearshore fish on small artificial reefs in Barkley Sound, British Columbia. Canadian Journal of Zoology 59:1635-1646.

Gerber, J. M. 1987. Fish use of artificial reefs in the central basin of Lake Erie. Master's Thesis, Ohio State University, Columbus.

Gilliam, J. F. 1982. Habitat use and competitive bottlenecks in size–structured fish populations. Ph.D. Dissertation. Michigan State University, East Lansing.

Gorham, J. C., and W. S. Alevizon. 1989. Habitat complexity and the abundance of juvenile fishes residing on small scale artificial reefs. Bulletin of Marine Science 44:662–665.

Grove, R. S., and C. J. Sonu. 1985. Fishing reef planning in Japan. Pp. 187–251 *in* F. D'Itri, editor. Artificial reefs: Marine and freshwater applications. Lewis Publishers, Inc., Chelsea, Michigan.

Grove, R. S., C. J. Sonu, and M. Nakamura. 1989. Recent Japanese trends in fishing reef design and planning. Bulletin of Marine Science 44:984–996.

Hairston, N. G., Sr. 1989. Ecological experiments: Purpose, design and exectution. Cambridge Studies in Ecology, Cambridge University Press, New York.

Hay, M. E., and J. P. Sutherland. 1988. The ecology of rubble structures of the South Atlantic Bight: A community profile, Biological Report 85(7.20). U.S. Department of the Interior, Fish and Wildlife Service, National Wetlands Research Center, Washington, D.C.

Helfman, G. S. 1979. Fish attraction to floating objects in lakes. Pp. 49–57 *in* D. L. Johnson and R. A. Stein, editors. Response of fish to habitat structure in standing water, Special Publication 6. American Fisheries Society, Bethesda, Maryland.

Herdendorf, C. E. 1985. Physical and limnological characteristics of natural spawning reefs in western Lake Erie. Pp. 149–183 *in* F. D'Itri, editor. Artificial reefs: Marine and freshwater applications. Lewis Publishers, Inc., Chelsea, Michigan.

Hixon, M. A., and J. P. Beets. 1989. Shelter characteristics and Caribbean fish assemblages: Experiments with artificial reefs. Bulletin of Marine Science 44:666–680.

Hixon, M. A., and W. N. Brostoff. 1985. Substrate characteristics, fish grazing, and epibenthic assemblages off Hawaii. Bulletin of Marine Science 37:200–213.

Holbrook, S. J., R. J. Schmitt, and R. F. Ambrose. 1990. Biogenic habitat structure and characteristics of temperate reef fish assemblages. Australian Journal of Ecology 15:489–503.

Hueckel, G. J., and R. L. Stayton. 1982. Fish foraging on an artificial reef in Puget Sound, Washington. Marine Fisheries Review 44:38–44.

Hueckel, G. J., R. M. Buckley, and B. L. Benson. 1989. Mitigating rocky habitat loss using artificial reefs. Bulletin of Marine Science 44:913–922.

Johnson, D. L., R. A. Beaumier, and W. E. Lynch, Jr. 1988. Selection of habitat structure interstice size by bluegills and largemouth bass in ponds. Transactions of the American Fisheries Society 117:171–179.

Kakimoto, H. 1982. The stomach contents of species of fish caught in artificial reefs. Pp. 271–273 in S. F. Vik, editor. Japanese artificial reef technology: Translations of selected recent Japanese literature and an evaluation of potential applications in the United States, Technical Report 604. Aquabio, Inc., Belleair Bluffs, Florida.

Kanayama, R. K., and E. W. Onizuka. 1973. Artificial reefs in Hawaii, Fish and Game Report 73-01. Hawaii Department of Land and Natural Resources, Honolulu.

Klima, E. F., and D. A. Wickham. 1971. Attraction of coastal pelagic fishes with artificial structures. Transactions of the American Fisheries Society 100:86–99.

Kock, R. L. 1982. Patterns of abundance variation in reef fishes near an artificial reef at Guam. Environmental Biology of Fishes 7:121–136.

Lehman, J. T. 1988. Ecological principles affecting community structure and secondary production by zooplankton in marine and freshwater environments. Limnology and Oceanography 33:931–945.

Lindquist, D. G., and L. J. Pietrafesa. 1989. Current vortices and fish aggregations: The current field and associated fishes around a tugboat wreck in Onslow Bay, North Carolina. Bulletin of Marine Science 44:533–444.

Lindquist, D. G., I. E. Clavijo, L. B. Cahoon, S. K. Bolden, and S. W. Burk. 1989. Quantitative diver visual surveys of innershelf natural and artificial reefs in Onslow Bay, N.C.: Preliminary results for 1988 and 1989. Pp. 219–227 in M. A. Lang and W. C. Jaap, editors. Diving for Science 1989. American Academy of Underwater Sciences, Costa Mesa, California.

Liston, C. R., D. C. Brazo, J. R. Bohr, and J. A. Gulvas. 1985. Abundance and composition of Lake Michigan fishes near rock jetties and a breakwater, with comparisons to fishes in nearby natural habitats. Pp. 492–514 in F. D'Itri, editor. Artificial reefs: Marine and freshwater applications. Lewis Publishers, Inc., Chelsea, Michigan.

Lopez, G. R. 1988. Comparative ecology of the macrofauna of freshwater and marine muds. Limnology and Oceanography 33:946–962.

Lowe-McConnel, R. H. 1987. Ecological studies in tropical fish communities. Cambridge University Press, New York.

Lukens, R. R. 1981. Ichthyofaunal colonization of a new artificial reef in the northern Gulf of Mexico. Gulf Research Reports 7:41–46.

Lynch, W. E., Jr., and D. L. Johnson. 1983. The effect of structure on fish behavior, Final Report, Project F-57-R. Ohio Department of Natural Resources, Federal Aid in Fish Restoration, Columbus.

Lynch, W. E., Jr., and D. L. Johnson. 1989. Influence of interstice size, shade, and predators on use of artificial structures by bluegills. North American Journal of Fisheries Management 9:219–225.

MacArthur, R. H., and E. O. Wilson. 1967. The theory of island biogeography. Princeton University Press, Princeton, New Jersey.

Mapstone, B. D., and A. J. Fowler. 1988. Recruitment and the structure of assemblages of fish on coral reefs. Trends in Ecology and Evolution 3:72–77.

Mathews, H. H. 1983. Primary production of artificial reefs in Florida waters. Third International

Artificial Reef Conference, Program and Abstracts, Newport Beach, California. (Abstract No. 9).

Matthews, K. R. 1985. Species similarity and movement of fishes on natural and artificial reefs in Monterey Bay, California. Bulletin of Marine Science 37:252–270.

McGurrin, J., and ASMFC Artificial Reef Committee. 1988. A profile of Atlantic artificial reef development, Special Report 14. Atlantic States Marine Fisheries Commission, Washington, D.C.

McGurrin, J., and ASMFC Artificial Reef Committee. 1989a. An assessment of Atlantic artificial reef development. Fisheries 14(4):19–25.

McGurrin, J. M., R. B. Stone, and R. J. Sousa. 1989b. Profiling United States artificial reef development. Bulletin of Marine Science 44:1004–1013.

McVey, J. P. 1970. Fishery ecology of the Pokai artificial reef. Ph.D. Dissertation, University of Hawaii, Honolulu.

Meier, M. H., R. Buckley, and J. J. Polovina. 1989. A debate on responsible artificial reef development. Bulletin of Marine Science 44:1051–1057.

Menge, B. A., and J. P. Sutherland. 1976. Species diversity gradients: synthesis of the roles of predation, competition, and temporal heterogeneity. American Naturalist 110:351–369.

Menge, B. A., and J. P. Sutherland. 1987. Community regulation: Variation in disturbance, competition, and predation in relation to environmental stress and recruitment. American Naturalist 130:730–757.

Milon, J. W. 1989. Economic evaluation of artificial reef habitat for fisheries: progress and challenges. Bulletin of Marine Science 44:831–843.

Molles, M. C. 1978. Fish species diversity on model and natural patch reefs: Experimental insular biogeography. Ecological Monographs 48:289–305.

Moring, J. R., M. T. Negus, R. D. McCulough, and S. W. Herke. 1989. Large concentrations of submerged pulpwood logs as fish attraction structures in a reservoir. Bulletin of Marine Science 44:609–615.

Mottet, M. G. 1985. Enhancement of the marine environment for fisheries and aquaculture in Japan. Pp. 13–112 in F. D'Itri, editor. Artificial reefs: Marine and freshwater applications. Lewis Publishers, Inc., Chelsea, Michigan.

Nakamura, M. 1985. Evolution of artificial fishing reef concepts in Japan. Bulletin of Marine Science 37:271–278.

Nolan, R. S. 1975. The ecology of patch reef fishes. Ph.D. Dissertation, University of California, San Diego.

Ogawa, Y. 1982. The present status and future prospects of artificial reefs: Development trends of artificial reef units. Pp. 23–41 in S.F. Vik, editor. Japanese artificial reef technology: Translations of selected recent Japanese literature and an evaluation of potential applications in the United States, Technical Report 604. Aquabio, Inc., Belleair Bluffs, Florida.

Ogden, J. C., and N. S. Buckman. 1973. Movements, foraging groups, and diurnal migrations of the striped parrotfish Scarus croicensis Bloch (Scaridae). Ecology 54:589–596.

Ogden, J. C., and J. P. Ebersole. 1981. Scale and community structure of coral reef fishes: A long-term study of a large artificial reef. Marine Ecology Progress Series 4:97–103.

Osman, R. W. 1978. The influence of seasonality and stability on the species equilibrium. Ecology 59:383–399.

Pardue, G. B. 1973. Production response of the bluegill sunfish, Lepomis macrochirus Rafinesque, to added attachment surface for fish food organisms. Transactions of the American Fisheries Society 102:622–626.

Pardue, G. B., and L. A. Nielsen. 1979. Invertebrate biomass and fish production in ponds with added attachment surface. Pp. 34–36 in D. L. Johnson and R. A. Stein, editors.

Response of fish to habitat structure in standing water, Special Publication 6. American Fisheries Society, Bethesda, Maryland.

Parker, R. O., Jr., D. R. Colby, and T. D. Willis. 1983. Estimated amount of reef habitat on a portion of the U.S. South Atlantic and Gulf of Mexico continental shelf. Bulletin of Marine Science 33:935–940.

Patriarche, M. H. 1959. Progress report on the effect of the Hoad fish shelter on fishing success. Michigan Department of Conservation, University Museums Annex, Ann Arbor.

Patton, M. L., R. S. Grove, and R. F. Harman. 1985. What do natural reefs tell us about designing artificial reefs in southern California? Bulletin of Marine Science 37:279–298.

Pettit, G.D., III. 1973. Stake beds as crappie concentrators. Proceedings of the Annual Conference of Southeastern Association of Game and Fish Commissioners 26:401–406.

Pierce, B. E., and G. R. Hooper. 1980. Fish standing crop comparisons of tire and brush fish attractors in Barkley Lake, Kentucky. Proceedings of the Annual Conference of Southeastern Association of Game and Fish Commissioners 33:688–691.

Polovina, J. J. 1989. Artificial reefs: Nothing more than benthic fish aggregators. California Cooperative Fisheries Investigations, CalCOFI Report 30:32–37.

Polovina, J. J., and I. Sakai. 1989. Impacts of artificial reefs on fishery production in Shimamaki, Japan. Bulletin of Marine Science 44:997–1003.

Prince, E. D., O. E. Maughan, and P. Brouha. 1985. Summary and update of the Smith Mountain Lake Artificial Reef Project. Pp. 401–430 in F. D'Itri, editor. Artificial reefs: Marine and freshwater applications. Lewis Publishers, Inc., Chelsea, Michigan.

Randall, J. E. 1963. An analysis of the fish populations of artificial and natural reefs in the Virgin Islands. Caribbean Journal of Science 3:31–46.

Reggio, V. C., Jr., compiler. 1989. Petroleum structures as artificial reefs: A compendium. Fourth International Conference on Artificial Habitats for Fisheries, Rigs-to-Reefs Special Session, November 4, 1987, Miami, Florida, OCS Study MMS 89-0021. U.S. Minerals Management Service New Orleans, Louisiana.

Relini, G., and L. O. Relini. 1989. Artificial reefs in the Ligurian Sea (Northwestern Mediterranean): Aims and results. Bulletin of Marine Science 44:743–751.

Rezak, R., T. J. Bright, and D. W. McGrail. 1985. Reefs and banks of the northwestern Gulf of Mexico. Wiley, New York.

Rice, D. W., T. A. Dean, F. R. Jacobsen, and A. M. Barnett. 1989. Transplanting of giant kelp Macrocystis pyrifera in Los Angeles Harbor and productivity of the kelp population. Bulletin of Marine Science 44:1070.

Richards, W. J., and K. C. Lindeman. 1987. Recruitment dynamics of reef fishes: Planktonic processes, settlement, and demersal ecologies, and fishery analysis. Bulletin of Marine Science 41:392–410.

Ricklefs, R. E. 1973. Ecology. Chiron Press, Portland, Oregon.

Rountree, R. A. 1989. Association of fishes with fish aggregation devices: Effects of structure size on fish abundance. Bulletin of Marine Science 44:950–959.

Russ, G. R. 1980. Effects of predation by fishes, competition, and structural complexity of the substratum on the establishment of a marine epifaunal community. Journal of Experimental Marine Biology and Ecology 42:55–69.

Russell, B. C. 1975. The development and dynamics of a small artificial reef community. Helgolander Wissenschaftliche Meeresuntersuchungen 27:298–312.

Russell, B. C., F. H. Talbot, and S. Domm. 1974. Patterns of colonization of artificial reefs by coral reef fishes. Proceedings of the Second International Coral Reef Symposium 1:207–215.

Rutecki, T. L., J. A. Dorr, III, and D. J. Jude. 1985. Preliminary analysis of colonization and succession of selected algae, invertebrates, and fish on two artificial reefs in inshore southeastern Lake Michigan. Pp. 459–489 *in* F. D'Itri, editor. Artificial reefs: Marine and freshwater applications. Lewis Publishers, Inc., Chelsea, Michigan.

Sale, P. F. 1977. Maintenance of high diversity in coral reef communities. American Naturalist 111:337–359.

Sale, P. F. 1980. The ecology of fishes on coral reefs. Annual Review of Oceanography and Marine Biology 18:367–421.

Sale, P. F. 1988. Perception, pattern, chance, and the structure of reef fish communities. Environmental Biology of Fishes 21:3–15.

Sale, P. F., and D. M. Williams. 1982. The ecology of coral reef fishes: Are the patterns more than expected by chance? American Naturalist 120:121–127.

Samples, K. C., and J. T. Sproul. 1985. Fish aggregation devices and open-access commercial fisheries: A theoretical inquiry. Bulletin of Marine Science 37:305–317.

Sato, O. 1985. Scientific rationales for fishing reef design. Bulletin of Marine Science 37:329–335.

Scarborough-Bull, A. 1989. Some comparisons between communities beneath petroleum platforms off California and in the Gulf of Mexico. Pp. 47–50 *in* V. C. Reggio, Jr., compiler. Petroleum structures as artificial reefs: A compendium, OCS Study MMS 89-0021. U.S. Minerals Management Service, New Orleans, Louisiana.

Schoener, A. 1982. Artificial substrates in marine environments. Pp. 1–22 *in* J. Cairns, Jr., editor. Artificial substrates. Ann Arbor Science Publishers, Ann Arbor, Michigan.

Schroeder, R. E. 1987. Effects of patch reef size and isolation on coral reef fish recruitment. Bulletin of Marine Science 41:441–451.

Seaman, W., Jr., R. M. Buckley, and J. J. Polovina. 1989. Advances in knowledge and priorities for research, technology and management related to artificial aquatic habitats. Bulletin of Marine Science 44:527–1073.

Shinn, E. A. 1974. Oil structures as artificial reefs. Pp. 91–96 *in* L. Colunga and R. B. Stone, editors. Proceedings of an International Artificial Reef Conference, TAMU-SG-74-103. Texas A&M University Sea Grant College Program, College Station, Texas.

Shinn, E. A., and R. I. Wicklund. 1989. Artificial reef observations from a manned submersible off southeast Florida. Bulletin of Marine Science 44:1051–1057.

Shulman, M. J. 1984. Resource limitation and recruitment patterns in a coral reef assemblage. Journal of Experimental Marine Biology and Ecology 74:85–109.

Shulman, M. J. 1985. Recruitment of coral reef fishes: Effects of distribution of predators and shelter. Ecology 66:1056–1066.

Shulman, M. J., and J. C. Ogden. 1987. What controls tropical reef fish populations: Recruitment or benthic mortality? An example in the Caribbean reef fish *Haemulon flavolineatum*. Marine Ecology Progress Series 39:233–242.

Shulman, M. J., J. C. Ogden, J. P. Ebersole, W. N. McFarland, S. L. Miller, and N. G. Wolf. 1983. Priority effects in the recruitment of juvenile coral reef fishes. Ecology 64:1508–1513.

Smith, B. W., G. R. Hooper, and C. S. Lawson. 1981. Observations of fish attraction to improved artificial mid-water structures in freshwater. Proceedings of the Annual Conference of the Southeastern Association of Fish and Wildlife Agencies 34:404–409.

Smith, C. L. 1977. Coral reef fish communities—order and chaos. Proceedings of the Third International Coral Reef Symposium 1:xxi–xxii.

Smith, C. L. 1978. Coral reef communities: A compromise view. Environmental Biology of Fishes 3:109–128.

Smith, G. B., D. A. Hensley, and H. H. Mathews. 1979. Comparative efficacy of artificial and natural Gulf of Mexico reefs as fish attractants. Florida Marine Research Publications 35:1–7.

Solonsky, A. C. 1985. Fish colonization and the effect of fishing activities on two artificial reefs in Monterey Bay, California. Bulletin of Marine Science 37:336–347.
Stephan, C. D., and D. G. Lindquist. 1989. A comparative analysis of the fish assemblages associated with old and new shipwrecks and fish aggregragating devices in Onslow Bay, North Carolina. Bulletin of Marine Science 44:567–579.
Stevenson, J. C. 1988. Comparative ecology of submerged grass beds in freshwater, estuarine, and marine environments. Limnology and Oceanography 33:867–893.
Stone, R. B., compiler. 1985. National artificial reef plan, NOAA Technical Memorandum NMFS OF-6. U.S. Department of Commerce, Washington, D.C.
Stone, R. B., H. L. Pratt, R. O. Parker, Jr., and G. E. Davis. 1979. A comparison of fish populations on an artificial and natural reef in the Florida Keys. Marine Fisheries Review 41:1–11.
Sutherland, J. P. 1974. Multiple stable points in natural communities. American Naturalist 108:859–873.
Sutherland, J. P., and R. H. Carlson. 1977. Development and stability of the fouling community at Beaufort, North Carolina. Ecological Monographs 47:425–446.
Talbot, F. H., B. C. Russell, and G. R. V. Anderson. 1978. Coral reef fish communities: Unstable high-diversity systems? Ecological Monographs 48:425–440.
Turner, C. H., E. E. Ebert, and R. R. Given. 1969. Man made reef ecology. California Department Fish and Game Bulletin 146:1–221.
Underwood, A. J., and P. G. Fairweather. 1989. Supply-side ecology and benthic marine assemblages. Trends in Ecology and Evolution 4:16–20.
Vance, R. R. 1988. Ecological succession and the climax community on a marine subtidal rock wall. Marine Ecology Progress Series 48:125–136.
Walsh, W. J. 1985. Reef fish community dynamics on small artificial reefs: The influence of isolation, habitat structure, and biogeography. Bulletin of Marine Science 36:357–376.
Walton, J. M. 1979. The effects of an artificial reef on resident flatfish populations. Marine Fisheries Review 44:45–48.
Warner, R. R., and T. P. Hughes. 1988. Population dynamics of reef fishes. Proceedings of the Sixth International Coral Reef Symposium 1:149–155.
Wendt, P. H., D. M. Knott, and R. F. Van Dolah. 1989. Community structure of the sessile biota on five artificial reefs of different ages. Bulletin of Marine Science 44:1106–1122.
Werner, E. E. 1986. Species interactions in freshwater fish communities. Pp. 231–245 in J. Diamond and T. J. Case, editors. Community ecology. Harper & Row, New York.
Wiebe, W. J. 1988. Coral reef energetics. Pp. 231–245 in L. R. Pomeroy and J. J. Alberts, editors. Concepts of ecosystem ecology. Springer-Verlag, New York.
Wilson, T. C., and R. E. Schlotterbeck. 1989. Assessment of rockfish utilization at the San Luis Obispo County artificial reef. Bulletin of Marine Science 44:1073.
Woodhead, P. M. J., and M. E. Jacobson. 1985. Epifaunal settlement, the process of community development and succession over two years on an artificial reef in the New York Bight. Bulletin of Marine Science 37:364–376.
Woodhead, P. M. J., J. H. Parker, and I. W. Duedall. 1985. The use of by-products from coal compustion for artificial reef construction. Pp. 265–292 in F. D'Itri, editor. Artificial reefs: Marine and freshwater applications. Lewis Publishers, Inc., Chelsea, Michigan.

Design and Engineering of Manufactured Habitats for Fisheries Enhancement

R. S. GROVE
Environmental Affairs
Southern California Edison Co.
Rosemead, California

C. J. SONU
Tekmarine, Inc.
Pasadena, California

M. NAKAMURA
Tokyo University of Fisheries
Tokyo, Japan

I. Introduction

Modern understanding of the engineering of artificial aquatic habitats has its origins in a diverse history of observations of fishes interacting with submerged objects. However, design, construction, and deployment of fisheries habitat is still a developing technology. It only recently has merged with and adapted to such ocean engineering practices as shoreline protection, harbor design, ship mooring, and marine pipeline technology.

From the standpoint of materials and deployment, a global review of artificial habitats reveals a broad array of technologies. In many areas, traditional practices of manual assembly of natural materials in small-scale structures remain popular; elsewhere, a contrasting but still low-cost approach in recent decades has been to deploy surplus materials over both small and large tracts. By contrast, in a few areas there has evolved a more comprehensive approach that applies engineering principles to all phases of habitat planning, from factory design and fabrication to on-site deployment

Artificial Habitats for Marine and Freshwater Fisheries
Copyright © 1991 by Academic Press, Inc.
All rights of reproduction in any form reserved.

of the largest structures used so far to enhance fishery habitat. At this stage of its evolution in fishery science, when circumstances warrant, sound engineering practice makes possible the construction and deployment of structures at sea that are equivalent volumetrically to medium-size multi-story apartment buildings.

This chapter provides a brief summary, from an engineering viewpoint, of the practices that have been used in freshwater, estuarine, and ocean environments to enhance fisheries. Emphasis is on relatively large structures, as opposed to small and easily hand-made devices such as those used to modify local segments of streams or shallow artisanal fishing grounds (Fig. 4.1). The principal focus is on those aspects of design, construction, and siting that are based on analytical research and use a quantitative approach consistent with currently accepted engineering practice in other fields. In some respects, this subject marks a new frontier for the integration of biology and engineering, as described in section VII on marine ranching.

A. Genesis of Building Practices

Fisheries enhancement using man-made devices and structures has found its way historically through the world as a logical extension of fishing management. Submerged objects such as fallen trees and cut brush, as well as sunken ships and intentionally placed rocks, were recorded to have aggregated fish as early as the late 1600s in Japanese coastal waters (Mottet, 1985), in the mid-1800s in United States estuarine waters (Stone, 1985), 1930s in U.S. lakes (Rutecki *et al.*, 1985), and early 1970s in U.S. reservoirs (Prince *et al.*, 1985). In Japan, with its heavy reliance on commercial fisheries, artificial habitat development became an active component of fisheries management in the 1930s (Mottet, 1981, 1985) and developed into a significant national program by 1954 (Grove and Sonu, 1985). The United States, Italy, and Taiwan were not far behind Japan: In the United States, as early as the 1950s the states of Alabama (Futch, 1981), California (Duffy, 1974), and Florida (Seaman and Aska, 1985) had built reefs. Italy began its efforts in 1970 (Bombace, 1989; Relini and Relini, 1989). Taiwan built its first concrete block reefs in 1957, and its national support program began in 1973 (Chang, 1985). Chapters 1 and 2 provide a review of these and other efforts, with illustrations.

In terms of scope and extent, efforts outside of Japan have not matched those of this island nation, which by 1970 had deployed concrete block artificial reefs in 3866 coastal locations (Ino, 1974). Further, between 1975 and 1982, 2200 additional reefs were placed in Japan using national annual subsidies equivalent to U.S. $50 million (Mottet, 1985). Thus, much of the dis-

Figure 4.1 Contrasting practices of artificial habitat design and construction (both effective). (A) Small-scale, traditional (i.e., artisanal) fishermen in Mexico construct a lobster shelter of logs (source: U.S. National Research Council, 1988). (B) Commercial fishing in Japan uses large benthic structures manufactured in factories.

cussion of design in this chapter is based on the Japanese experience, which heretofore has not been completely available due to limited translation.

1. Trend toward Design Manuals

Design of large fishing reefs long remained an art, a practice guided largely by the trials and errors of individuals. Beginning in the late 1960s, a trend toward increased objective reasoning reflects: (1) accumulation of knowledge on fish behavior and its response to man-made structure, and (2) advances in the scientific understanding of physical processes in the coastal ocean. Merging the growing knowledge of fish behavior and physical environment gave reef designers a more rational approach to seeking optimum harmony between fishes, ambient physical conditions, and structure. Again, this has been most apparent in Japan (Sato, 1985).

This trend began when humans, wishing to take advantage of certain known behaviors of fish, began to fabricate artificial reefs with new materials such as concrete, steel, fiber-reinforced plastics, and fly ash, instead of merely deploying rock, discarded car bodies, sunken vessels and lumber. When fabricating a reef, adequate shapes and dimensions that have desirable structural integrity and durability must be ascertained based on knowledge of the environmental loading anticipated, as well as strength of the material to be used. Much of this chapter is devoted to these aspects of the subject.

Japan is the only country where reef design processes have been codified into official manuals. The first comprehensive design manual, *Coastal Fisheries Development Program: Structural Design Guide* (359 pages) was prepared in 1978 and extensively revised in 1984 (Japan Coastal Fisheries Promotion Association [JCFPA], 1984). Subsequent JCFPA manuals address steel reefs (98 pages; 1982), planning aspects (1184 pages; 1986), and design (398 pages; 1989). They were prepared under the auspices of the quasi-governmental JCFPA, and set forth the criteria by which all new reef projects are judged for eligibility for Japanese government subsidy (Grove *et al.*, 1989). Although the manuals are based on considerable experience, they are far from complete.

The knowledge of coastal engineering and nearshore oceanography is yet to be assimilated into reef planning and design procedures in sufficient detail and scope. For instance, many of the methods specified in the Japanese manuals for estimating wave and current forces do not go much beyond the long-established American Petroleum Institute (API) design codes for offshore structures and marine pipelines, nor the standard design procedures for coastal facilities described in the *U.S. Army Corps of Engineers Shore Protection Manual* (U.S. Army Corps of Engineers, 1984), and widely regarded as one of the most advanced textbooks on coastal engineering. Be-

cause of the low tolerance for failure of these sensitive structures, procedures following the API codes or the *Shore Protection Manual* could result
in overdesign for habitat structures.

II. Structures Used as Aquatic Habitats

A wide array of materials have been used to construct artificial habitats (see
Fig. 1.1). The most primitive is rock. Rocks have been placed either singly,
as a pile, in wooden cribs, or in scuttled boats.

Man-made materials used for benthic structures include concrete, iron
and steel, reinforced concrete (concrete and iron), ceramic, plastic, plastic concrete (concrete mixed with polyethylene, polypropylene, sand, and
iron), fiber-reinforced plastic (FRP), and asbestos fiber, among others (Sonu
and Grove, 1985). These materials are used in structures that are fabricated on land according to particular design specifications. However, this
approach is far from universal.

Materials of opportunity still account for a large part of reef construction. Surplus and scrap materials are often obtained at no cost and deployed
without assembly or significant modification, except for cleaning to eliminate
environmental hazards. Materials include derelict ships, automobile bodies,
automobile tires, debris from demolition projects, and even discarded offshore oil platforms. Recently, waste combustion byproducts from fossil fuel-
fired electricity generating plants (i.e., mixes of fly ash with flue-gas de-
sulfurization scrubber sludge) have been experimentally tested (Woodhead
et al., 1985). Use of old offshore platforms to convert drilling wastes, garbage, and shipboard refuse into environmentally safe reef building blocks
also has received evaluation (Anonymous, 1989).

Older applications in so-called artisanal (or traditional) fisheries use
natural materials that include not only rock but also brush piles and log cribs
on the bottom, and floating rafts of bamboo, coconut fronds, and cork.

A. Categories of Applications

The many varieties of artificial habitats and components can be catego-
rized into major groups according to intended use and design. Tables 4.1
and 4.2 review artificial reefs and habitats for three primary environments
(freshwater, estuarine, and ocean) according to intended environmental or
ecological function. Despite the lack of a systematic global data base for artificial habitats (see Chapters 1 and 2), it is clear that the greatest variety of
materials and applications have been in the ocean. Representative structures
and biological impacts are described in Chapter 5.

TABLE 4.1
Principal Types of Natural and Man-made Materials Used in
Artificial Habitats in the Aquatic Environment

Material and structure	Environment and application[a]		
	Ocean	Estuary	Fresh-water
Natural materials			
Bamboo	C	—	—
Brush	A	—	—
Coconut	A	—	—
Oyster shell	—	H	—
Quarry rock	R,H,M,E	R	R
Rope	A	—	—
Stone (piled or in gabions)	H	H	R
Trees, logs	H	—	R
Wooden frames	R	R	—
Manufactured or scrap products			
Concrete			
Poured structures	R,C,H,E	R,H,E	E
Rubble	R,H	R	R
Fiberglass/plastic			
Benthic reef modules	R,C	—	—
Midwater buoys, streamers	R,C,A,E	R,H	—
Seaweed	H	—	R
Incineration ash	E	—	E
Rubber			
Automobile tires	R,C,A,H,E	R,H	R
Steel			
Automobile bodies	R,C,H	R,H	—
Benthic reef modules	C	—	—
Fuel storage tanks	R	—	—
Petroleum production			
platforms	R	—	—
Street cars (trolleys)	R	—	—
Vessels	R,H	R	—
Wood			
Vessels	C	—	—

[a] Abbreviations in columns indicate, in descending order of relative importance, principal use of structure: A, artisanal (small-scale) fishing; C, commercial fishing; E, experimental; H, habitat enhancement; M, mitigation; R, recreational fishing.

TABLE 4.2
Classification of Larger Artificial Habitat Structures by Environments

System	Water body	Intended use
Freshwater	Lakes and reservoirs	Reefs of opportunity (breakwaters, jetties, industrial structures such as water intakes) Fish-aggregating devices (FADs) Enhancement and aggregation reefs Nursery grounds
	Rivers	Reefs of opportunity (river bank revetment, groins) Spawning and nursery area improvement
Estuarine	Inlets and sounds	Shore protection structures Enhancement and aggregation reefs
	Bays and canals	Breakwaters Nursery and spawning grounds Enhancement and aggregation reefs Marine ranches
Ocean	Shallow water (0–20 m depth)	Reefs of opportunity (jetties, groins, breakwaters, submerged intakes and outfalls, oil production platforms) Enhancement reefs for plants, fish and invertebrates Fish-aggregation reefs Underwater obstacles to discourage excessive fish trawling Marine ranches Fish-aggregating devices
	Offshore (20 m and deeper)	Operational oil platforms Dissembled oil platforms Migratory fish-attraction reefs Fish and invertebrate enhancement reefs Fish-aggregating devices

1. Coastal Works and Shore Protection Structures

Traditional waterfront facilities such as breakwaters, jetties, groins, and various intake and outfall devices for power (Helvey and Smith, 1985) and sewage treatment stations have been known to enhance plant, invertebrate, and fish populations in freshwater (Binkowski, 1985) and estuarine (Alevras and Edwards, 1985; Burchmore *et al.*, 1985) environments. While these structures serve other primary purposes, they function in many instances as effective reefs (Table 4.2). Their construction is addressed in civil engineering manuals. Oil-drilling platforms, which penetrate the entire water column, are a particularly unique facility that can attract fish in all the layers of the water from the bottom to the surface (Quigel and Thornton, 1989).

2. Fish-Aggregating Devices

Observations of fish congregating around floating rafts and anchored buoys (Mottet, 1985; Bombace, 1989) led to the concept and refinement of surface and midwater fish-aggregating devices (FADs). Typical FADs manufactured from man-made materials consist of plastic or nylon streamers or mesh, typically framed in pieces of polyvinyl chloride (PVC) pipe, attached to line with an anchor at one end and a surface or midwater float at the other (Fig. 4.2). FADs of natural materials such as cork are used in many artisanal settings (see Figs. 1.1 and 1.4 in Chapter 1). Materials of opportunity used in FAD construction include concrete-filled automobile tires for anchors, foam-filled tires or bamboo for floats, and palm fronds for midwater streamers (see Figs. 5.1 and 5.2).

A review of the literature indicates that most oceanic FADs are short-lived, usually breaking away from their moorings within a matter of weeks to months (Buckley *et al.*, 1989). Some Japanese designs, however, have an estimated structural life of 15 yr (Mottet, 1985). The potential environmental impact of FAD components that may break away needs to be considered in planning. Hawaii and countries in the tropical Pacific, Atlantic, and Indian Oceans continue to refine FADs through use of stronger, more durable materials and better designs. Research on FADs is a major component in the Japanese marine ranching program (Agricultural, Forestry, and Fishery Technology Council [AFFTC], 1989).

3. Artificial Benthic Reefs

The goal of active manipulation of nursery areas and spawning grounds motivated empirical development of shallow water, low-profile artificial reefs for fishery resource enhancement (Mottet, 1985). Discovery and study of the thigmotropic nature of certain fishes and their affinity for, or orientation to, vertical structures or bathymetric relief in the environment enhanced this process. Such observations in shallower waters, where the first artificial reefs were placed (Grove and Sonu, 1985; Stone, 1985), would have contributed to design of more effective habitats. Larger, offshore fish-attraction reefs (up to 12 m in height) and the deeper placement of more typical 1–3 m high reef modules evolved from experimentation in aggregating commercially important, migratory oceanic fishes (Nakamura, 1985).

In this section, we review the types of prefabricated structures placed on the seafloor (but apparently not on lakebottoms) that must be subjected to more detailed engineering considerations due to the relatively great stresses of the physical environment. Design of smaller structures, such as the typical small-scale "fish attractors" used in streams and lakes (see Fig. 1.2), is addressed in references such as Phillips (1990). Meanwhile, various designs of more massive artificial reefs, especially those used in Japan, have

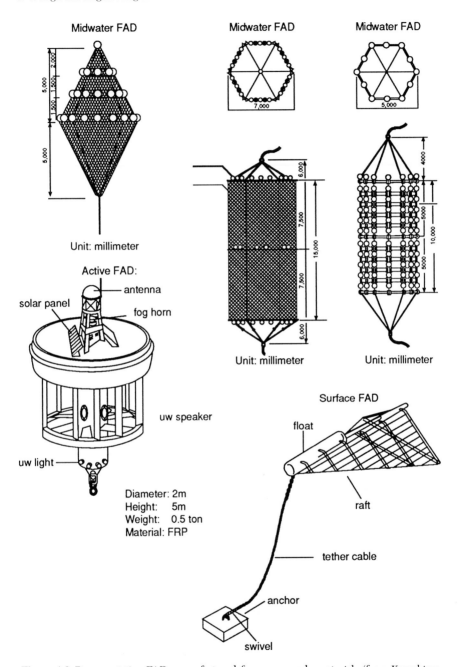

Figure 4.2 Representative FADs manufactured from man-made materials (from Kagoshima prefecture, 1987; Matsumi, 1987).

been reviewed by Mottet (1981) and Grove and Sonu (1985). The latter addresses 68 well-known varieties of artificial reefs from Japan, but many more
have emerged since then. The typical reef shapes exhibited in Figs. 4.3 and
4.4 reveal the size range of artificial habitats; see also Fig. 1.5. There are
literally hundreds of artificial reef designs around the world, although as
discussed later the materials used are fairly limited in number (Table 4.2).

In Japan, the simple designs such as cube and tube (Fig. 4.3) may be
called "blocks," as distinguished from more complicated designs such as
those incorporating beams, panels, and columns, which may be called "modules" (Fig. 4.4). The blocks are elementary components grouped to form
reefs. Modules typically are large enough to function as individual reefs.
(Floating structures also may be called modules because of their large fabricated design, although their appearance is conspicuously different from the
benthic modules.) Many Japanese commercial-fishing reef modules feature
unique shapes, often with highly intricate appearances. These shapes are
adopted to induce enhanced fish aggregation, as well as to improve structural integrity and safety of handling.

In the Mediterranean Sea, Italy and France have established that the
1–2 m hollow concrete cube, specifically designed for artificial habitats, is
the preferred shape, size, and material for reefs (Bombace, 1989; Fabi et al.,
1989; Relini and Relini, 1989). Their goals are multipurpose, including fish
aggregation, fish and shellfish shelter, resource enhancement, and also de-
terrence of illegal trawling. The blocks offer great versatility in reef design
and are either stacked into pyramids or scattered separately throughout the
fishing grounds or are used in a combination pattern of both pyramids and
scattered pattern. Concrete blocks of two sizes, 1 m^3 and 12 m^3, have been
found to be the most durable materials used in Taiwan (Chang and Shao,
1988).

Areas of the southeastern United States have begun to use and experiment with concrete cubes, modules, and other designs such as prefabricated
fiberglass-reinforced plastic (FRP) cylinders (Sheehy, 1983; Bell et al., 1989;
Sheehy and Vik, 1989). On the Pacific coast of the United States, the California Department of Fish and Game has determined that artificial reefs
should continue to be made from quarry rock. This was based on observations in the 1960 through 80s that although "concrete cubes" were most
effective in attracting fishes, quarry rock is a close second and is economically more practical (Lewis and McKee, 1989). More recent studies have
further substantiated the value of quarry rock, due to its greater potential
for colonization by and production of food organisms (Lewis and McKee,
1989). Clean concrete rubble (i.e., surplus) will continue to be an acceptable
material in California reef projects, since it has similar properties to quarry

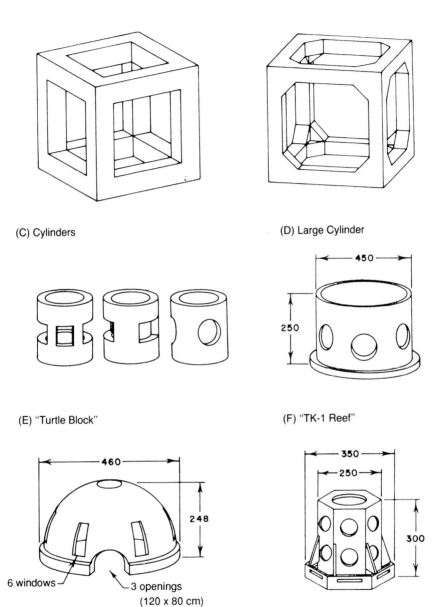

(A) Cube

(B) Cube

(C) Cylinders

(D) Large Cylinder

450

250

(E) "Turtle Block"

460

248

6 windows

3 openings
(120 x 80 cm)

(F) "TK-1 Reef"

350

250

300

Figure 4.3 Representative reef blocks: (A) Cube, (1 to 5 m on a side; 1 to 125 m³); (B) Cube, (1 to 4 m on a side; 1 to 64 m³); (C) Cylinders, (diameters and heights 0.6 to 1.8 m; 0.17 to 4.6 m³); (D) Large Cylinder, (39.76 m³); (E) "Turtle Block," (13 tons; 27.3 m³); (F) "TK-I Reef," (13.1 tons; 28.9 m³), which can be deployed separately or in groups. These can be made of concrete or fiberglass (Unit: mm) (From Mottet, 1981).

120

R. S. Grove, M. Nakamura, and C. J. Sonu

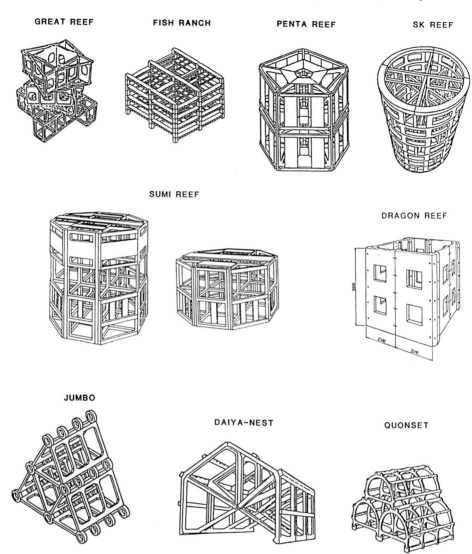

Figure 4.4 Representative reef modules (from Grove and Sonu, 1985). These benthic structures are considerably larger than the blocks in the preceding figure, and deployment is correspondingly more complex (see Fig. 1.5).

rock (Sheehy, 1983; K. Wilson, California Department of Fish and Game, personal communication, 1990).

To recapitulate, fishery interests have created habitats from both natural and man-made materials, and sometimes public works projects create habitat as a secondary result. Although materials of opportunity such as derelict vessels, discarded automobile tires, and various high-density, durable, surplus structures commonly are used in artificial habitats, the emphasis of this review and the following discussion is on the engineering aspects of fabrication and deployment of designed habitats.

B. Size Categories

Artificial habitats of different size may be constructed by deploying individual reef units in specially arranged groupings and organizations. In Japan, for example, reefs are classified into four categories depending upon the size of fishing grounds to be created. These are called, in ascending order of size, (1) reef unit, (2) reef set, (3) reef group, and (4) reef complex (Fig. 4.5).

In this hierarchy, proposed by Grove and Sonu (1983), the reef unit is the smallest element and is represented by individual reef "blocks" as described previously (Fig. 4.3). Used as modules, the reef units may be deployed as a reef set. According to Japanese experience (Nakamura, 1985), the typical reef set may be up to 300 m in radius, and to be viable the reef set must have a minimum 400 cm of reef bulk volume. Reef bulk volume, as used by Japanese reef experts to denote volumetric size of reef, represents a space enclosed by the exterior envelope of the reef, including both the reef bodies and the spaces between them.

Reef sets of different bulk volumes can be constructed by changing the deployment patterns for individual reef blocks. Namely, loose deployment of reef blocks will enhance the size of the envelope, but the reef's impact on fish may diminish if the blocks are scattered too widely. Japanese fishing reef planning guidelines (JCFPA, 1986) recommend that the deployment area of reef blocks should not exceed 20 times the aggregate shadow areas of all individual blocks. Mathematically, this rule is expressed as:

$$S < 20 \cdot N \cdot X \qquad (1)$$

in which S is the deployment area of reef blocks, N the number of reef blocks deployed, and X the shadow area of an average reef block.

The criterion shown in Eq. (1) is highly useful when, for instance, planning for adequate numbers of blocks N, adequate bulk volume J, and adequate radius of deployment r, which are required for a viable reef set. For instance, assume that a deployment area is circular with radius r, so that

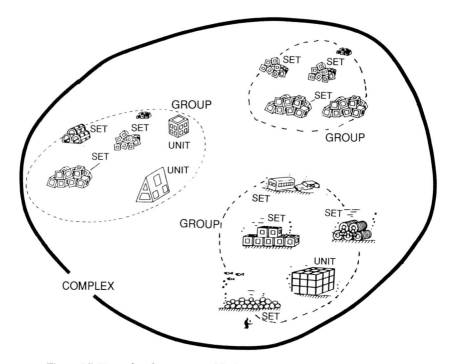

Figure 4.5 Hierarchy of Japanese reef deployment (from Grove and Sonu, 1983).

$S = \pi r^2$, and consider a cube block with a side length a, so that $X = a^2$. The bulk volume consisting of an N number of such blocks will be $J = a^3 N$, or, alternately, $N = J/a^3$. Using these relationships, Eq. (1) can be rewritten as:

$$r^2 \pi < 20 \cdot (J/a^3) \cdot (a^2)$$

or

$$r < \sqrt{20 \cdot (J/a)/\pi} \qquad (2)$$

Consider a reef set with a minimum required bulk volume of 400 cm ($J = 400$). The maximum radius of deployment area for this reef using blocks with a side length $a = 1.5$ m is, from Eq. (2), $r = 41$ m. A total of 120 blocks ($N = J/a^3 = 400/1.5^3$) are required to construct this reef.

Usually, more than one reef set is needed to construct a fishing ground large enough to support commercial fishing. For this purpose, reef sets may be assembled into the so-called reef group where the reef sets are deployed in such a way that areas of influence around individual reef sets are allowed to overlap with each other. Since the areas of influence have typical radii of 200 to 300 m, maximum separation between adjacent reef sets in a reef group should be between 400 and 600 m. Usually, a reef group is large enough to intercept a migratory fish path or even support a marine ranching operation. Figure 4.6 shows a diagrammatic example of an actual reef group.

The largest fishing ground that can be constructed with artificial reefs is the reef complex. The reef complex is formed by assembling more than one reef group into a coherent layout, usually placed in a string along the path of seasonal fish migration or water-mass movements. This arrangement allows fishing activities to move successively from one area to another extending the overall fishing season. In the reef complex, individual reef groups need not interact with each other. Thus, a single reef complex in its entirety may even encompass a whole geographical region.

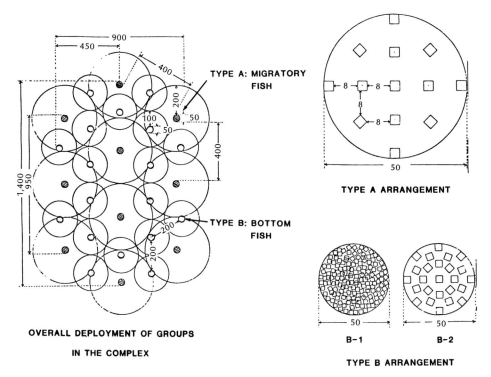

Figure 4.6 Horizontal configuration of a representative reef group in Japan (from Grove and Sonu, 1985). (All dimensions in meters.)

Environmental assessment and monitoring of reef complexes and their components (i.e., reef units, sets, and groups) needs to be considered in their design and planning. Ecological aspects are reviewed in Chapter 3.

III. Materials

The principal attributes for artificial reef materials are durability, safety, functionality, and economy. In Japan, for instance, the government has established standards for each attribute as a requirement for construction subsidies: for durability, the reef material must assure a minimum 30-yr life without deterioration; for safety, it must be free of toxic substances; for functionality, the material must have demonstrated fish-aggregation capability based on field verification testing for a minimum of one year; and for economy, it must be cost-effective. Concrete and steel generally meet these standards and are most commonly used as materials for artificial reefs. Other satisfactory materials include "polycon" (a mixture of cement, sand, iron, and waste polyethylene), and fiber-reinforced plastic (Grove and Sonu, 1985).

A. Safety Standards

Reefs must be made of materials with chemical properties that will not endanger workers during handling and placement or harm marine life while in the water. In general, environmental regulations in a number of countries strictly prohibit any PCB (polychlorinated biphenyl), mercury, cyanide, and organic phosphate in materials used for fishing reefs, and specify allowable contents of such hazardous materials as cadmium, lead, chrome, and arsenic.

In the United States, such regulations are provided in the federal Waste Disposal Act, as amended by the Resource Conservation and Recovery Act of 1976, and in various state regulations (for example, total or soluble threshold concentrations allowed for heavy metals and pesticides are regulated by Title 22, California Administrative Code, Article 11). According to these regulations, a waste is classified as hazardous in reference to four characteristics: ignitability, reactivity, corrosivity, and "EP toxicity." The EP (extraction procedure) toxicity relates to allowable maximum concentrations, to be determined by standard procedures specified by the U.S. Environmental Protection Agency, of the leachate for eight trace metals contained in the material: i.e., silver, arsenic, barium, cadmium, chrome, lead, mercury, and selenium. For a detailed discussion of the regulations see Parker *et al.* (1984).

B. Concrete

Concrete is one of the most widely used materials for fabrication of reef modules today. Types of cement used in construction include low-heat-of-hydration, Portland cement, sulfate-resisting cement, silica cement, and fly-ash cement. Although offering excellent long-term strength and durability, these cements are typically low in developing early strength, requiring special care during initial curing and installation.

Allowable maximum stresses may be specified for the design of concrete members of the reef. For instance, according to guidelines developed in Japan (JCFPA, 1984), the maximum compressive stress due to bending should not exceed $\sigma/4$ (or 55 kg/cm^2) and $\sigma/3$ (60 kg/cm^2) for nonreinforced and reinforced concrete members, respectively, where σ denotes the 28-day strength of the concrete under standard curing. Likewise, the maximum allowable tensile stress due to bending in nonreinforced concrete should not exceed $\sigma/3$ or 3 kg/cm^2.

Water–cement ratio is the principal factor affecting the strength of concrete. Table 4.3 summarizes the water–cement ratios recommended by the 1984 Japanese design manual. The reader should note that the offshore boundary of the surf zone is defined by a water depth equal to the height of the design wave for a given location.

TABLE 4.3
Recommended Water–Cement Ratios for Concrete Reefs[a]

	Mixture			
Site condition	Water–cement ratio (%)	Range of slump (cm)	Maximum size of coarse aggregate (mm)	Maximum 28-day strength (kg/cm^2)
Away from surf zone	60	3~8	40~50	160
Within and close to surf zone	55	3~8	40~50	180

[a] Source: JCFPA, 1984.

TABLE 4.4
Rate of Corrosion of Steel Plate in Seawater[a]

Side facing	Site conditions	Rate of corrosion (mm/year)
Offshore	Above high-water level	0.3
	Between high-water level and seafloor	0.1
	Buried in seafloor	0.03
Shoreward	Subaerial	0.1
	Underground (aboveground water level)	0.03
	Underground (belowground water level)	0.02

[a](Source: JCFPA, 1984).

C. Steel

Typical parameters of the steel and cast iron that are used for the fabrication of artificial reefs in Japan are as follows (Nakamura, 1980):

Young's Coefficient	2.1×10^6 kg/cm^2
Poisson Ratio	0.30
Specific Gravity	7.85
Compression Strength	$4900 \sim 6300$ kg/cm^2

The rate of corrosion in seawater is a function of the environmental conditions at the site as well as the quality of the material. Although the guidelines shown above in Table 4.4 are useful, designers of steel reefs should take into consideration site-specific or similar experiences as well.

The Japanese guidelines recommend that welding, rather than nut-and-bolt connections, be used to fabricate a steel reef module (JCFPA, 1984). Specific design guidelines for steel reef modules are available (JCFPA, 1982).

IV. Structural Integrity of Reef Blocks

Structural damage to reef blocks could occur when they are rolled on the seafloor by wave and current action or when they impact the seafloor during installation. Perhaps the largest physical load to the block will occur when it lands on the seafloor after a free-fall through the water column (Nakamura et al., 1975). A method to compute the impact load on a reef block as it lands on the seafloor by free-fall is described below, following the procedures recommended by Japanese planning guides (JCFPA, 1989). Once safely on the seafloor after free-fall installation, blocks are likely to retain their structural integrity against most other loading situations that might arise subsequently.

A. Calculation of Reef Block Strength

A free-falling object accelerates as it travels downward in the water column from the surface, and soon reaches a steady speed called terminal velocity, v_c, when the drag force balances the downward gravitational pull. The terminal fall velocity is expressed as

$$v_c = \sqrt{2gV(\sigma_o/\rho - 1)/(C_D A)} \tag{3}$$

in which g is the gravitational acceleration, V is the volume of the structural members of the falling object, A is the projected area of the falling object, σ_o the density of the object, ρ the density of the water, and C_D the drag coefficient.

The fall velocity of a reef block usually reaches a terminal value within about 10 m from the water surface. Thus, the speed of an object impacting a seafloor deeper than 10 m may be approximated by v_c. In shallower water, an intermediate fall velocity, v, corresponding to water depth, z, is given by

$$v = v_c \sqrt{1 - e^{-2Kz}} \tag{4}$$

in which

$$K = C_D A / 2V(\sigma_o/\rho + C_{MA})$$

where C_{MA} is the virtual mass coefficient.

Displacement of the seafloor, ε_o, caused by the impact of an object with a landing speed v_o, is obtained from

$$L\varepsilon_o^{n+1} - M\varepsilon_o - N = 0$$

$$
\begin{aligned}
L &= gK_R/[(n+1)w_o V] \\
M &= g[(\sigma_G/w_o) - 1] - C_D A v_o^2/(4V) \\
N &= [(\sigma_G/w_o) + C_{MA}](v_o^2/2)
\end{aligned}
\tag{5}
$$

in which w_o is the unit weight of the water, σ_G is the unit weight of the reef block, K_R is the coefficient of seafloor reaction, and n is an integer number (Nakamura, 1980). The seafloor displacement, ε_o, is computed asymptotically using the Newton's approximation method as follows:

$$\varepsilon_1 = (N/L)^{1/(n+1)}$$

$$\varepsilon_{r+1} = \varepsilon_r - [L\varepsilon_r^{n+1} - M\varepsilon_r - N]/[(n+1)L\varepsilon_r^n - M] \tag{6}$$

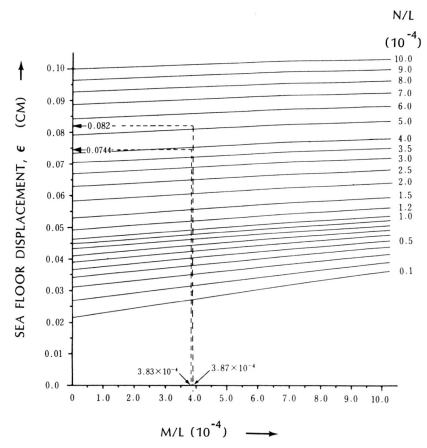

Figure 4.7 Nomogram showing relationship of L, M, N, and seafloor displacement, ε. See Eqs. (5) and (6) in the text. (Source: JCFPA, 1989.)

The value of ε_o also may be obtained using a nomogram as shown in Fig. 4.7. Maximum reaction of the seafloor, R_o, to the landing impact is thus:

$$R_o = K_R \varepsilon_o^n \qquad (7)$$

In order to simplify the computation of the required strength of the reef block, one may substitute the dynamic load due to the landing impact with a static load, σ_G', such that

$$\sigma_G' = R_o / V \qquad (8)$$

and

$$k = \sigma_G'/\sigma_G = (K_R \varepsilon_0^n)/(\sigma_G V) \qquad (9)$$

Thus, one essentially assumes that due to the landing impact, the unit weight σ_G has increased to σ_G' by a factor of k.

One may use $n = 2$, and $K_R = 500$ kg/cm^2 for sandy or gravel substrate, and $K_R = 630$ kg/cm^2 for a hard clay substrate. Typical values for drag coefficient and virtual mass coefficient are given in Table 4.5.

1. Example

The cube block shown in Fig. 4.8 measures 1.5 m × 1.5 m × 1.5 m in exterior dimensions with each window 1.0 m × 1.0 m. The volume of the structural members of the block is $V = 0.906$ m^3. The unit weights of the reinforced concrete and the seawater are

$$\sigma_G = 2.45 \text{ ton/m}^3, \quad \text{and} \quad w_o = 1.03 \text{ ton/m}^3$$

in which "ton" represents a metric ton (i.e., 1000 kg). Thus, one gets:

W (weight of the block) $= \sigma_G V = 2.22$ ton

W' (immersed weight of the block) $= (\sigma_G - w_o)V = 1.29$ ton

The ambient design conditions are as follows:

Design wave: Height 6.0 m, Period 12 sec
Design current:
    Speed $= U_{11} = 0.5$ m/sec at sea surface, and
    Direction $= 60°$ to wave direction
Depth of installation: 50 m
Method of installation: Free-drop from the surface
Substrate: Sandy (median diameter 0.5 mm)
Seabed reaction coefficient: $K_R = 5000$ ton/m^2

The projection area of the block during the free-fall is readily computed to be 2.27 m^2 when falling with the flat facedown, and 3.10 m^2 when falling at a tilt angle of 45° to the vertical.

Since the water depth is in excess of 10 m, the fall-speed at the time of landing essentially equals the terminal fall velocity, v_c. Thus, substituting into Eq. (3): $h = 50$ m, $V = 0.908$ m^3, $C_D = 2.0$, $C_{MA} = 1.0$, and $g = 9.8$ m/sec^2, one obtains $v_c = 2.32$ m/sec when falling facedown, and v_c

TABLE 4.5
Hydrodynamic Coefficients for Immersed Bodies

Shape	Box (current; a, b, c)	Cylinder (current; a, c)	Box (current; a, c)
c/a	1 2 4 5 10 20 25	1 2 5 10 20 40	1 2 4 5 10 20
C_D	1.05 1.08 1.13 1.14 1. 1.50 2.00	0.63 0.68 .074 0.82 0.90 1.00 -	1.12 1.15 1.19 1.20 1.29 1.50 2.00
C_{MA}	1.0	1.0	1.0
Area	ac	ac	ac
Volume	abc	$\pi a^2 c/4$	abc

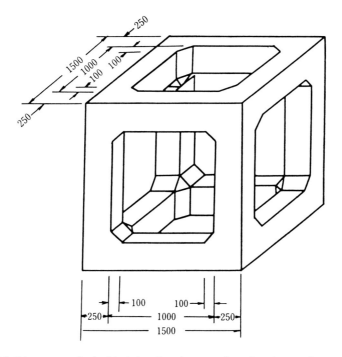

Figure 4.8 Dimensions of cube block (reinforced concrete) used in the sample computation. (Dimensions in millimeters.)

$= 1.99$ m when falling at a 45° angle. Use of Eq. (4) will easily verify that $v_c = v$ in both cases.

To compute the landing impact σ'_c, one must determine ε_o through L, M, and N in accordance with Eq. (5). In our case, $L = 17{,}502.88$, $M = 6.77$, and $N = 9.09$ for a facedown landing, and $L = 17{,}502.88$, $M = 6.74$, and $N = 6.70$ for an angular landing. To use the nomogram in Figure 4.7, from respective values of M/L and N/L, one gets $\varepsilon = 8.2 \times 10^{-2}$ (m) for a facedown landing and $\varepsilon = 7.44 \times 10^{-4}$ (m) for an angular landing. The corresponding impact-load expressed as equivalent unit static weight σ'_c is, from Eq. (8), 37.11 and 30.55 ton/m³, respectively. It then follows from Eq. (9) that the structural members of the block must be designed as if the unit weight of its material has increased by a factor of 15.15 and 12.47, respectively.

Figure 4.9 shows distribution of bending moment, axial stress, and shear stress on the structural members of the cube block due to the landing impact.

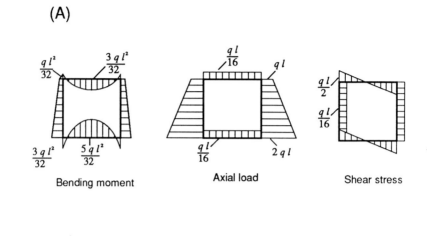

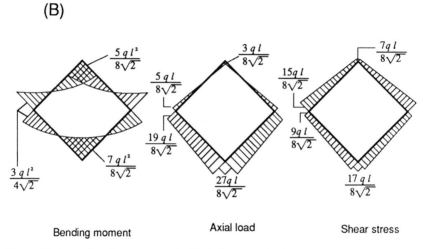

Figure 4.9 Stress distribution on a cube block due to the landing impact on the seafloor. (A) Face-down landing; (B) angular landing. (Source: JCFPA, 1984.)

V. Environmental Design Criteria

Effective design of large reef modules must consider prevention of failure of reef blocks due to wave and current actions. Modes of failure of particular interest are rupture of structural members, tilting and subsidence due to scour of its foundation, and sliding and toppling due to excessive lateral thrust of the ambient forces.

A. Design Wave

Procedures for deriving the design wave for a given location should follow the methods described in the Shore Protection Manual (U.S. Army Corps of Engineers, 1984). For most purposes, wave computations using small-amplitude theory are acceptable.

B. Design Current

Design current speed u_o may be derived from sea surface values, v_H, indicated on published navigational charts and tide tables, as follows (Nakamura, 1980):

$$u_o = k'v_H(h/D)^{1/7} \tag{10}$$

in which D is the water depth, and h is the height of the reef crest above the seafloor. This formula is based in part on a $\frac{1}{7}$th law of velocity profile at a solid boundary, a hypothetical simplification. A correction factor, k', is usually necessary to account for the complexity of real situations. Nakamura (1980), based on partial field verification, recommends the following values for k':

$$k' \begin{cases} = 1.6 \text{ in the direction paralleling the principal axis of the current} \\ = 1.2 \text{ in the direction normal to the principal axis} \\ = \sqrt{\cos \alpha + 1.5} \text{ in the direction at angle } \alpha \text{ to the principal axis} \end{cases}$$

It is recommended to derive v_H from actual measurements at the site. Nakamura (1980) recommends that v_H should be taken from a record not shorter than 15 days, either as a cumulative sum of all the harmonic amplitudes or as a maximum observed value in the record after applying a 12-hr running average.

C. Current Forces Acting on Reef Block

Forces acting on a reef block due to the combined action of wave and current, F, are expressed as follows:

$$F = F_D(\sin \theta + u_o/u_m)^2 - F_M\cos \theta \tag{11}$$

in which

$$F_D = C_D A(w_o/2g)u_m^2, \\ F_M = C_M V(w_o/g)(2\pi/T)u_m \tag{12}$$

in which C_D is the drag coefficient, $C_M = C_{MA} + 1$, C_{MA} is the virtual mass coefficient, A is the projection area of the structural members of the reef against the principal wave direction, V is the volume of the reef members, w_o is the unit weight of the water, g is the gravitational acceleration, θ is the phase angle ($= 2\pi x/L - 2\pi t/T$), T is the wave period, and u_m is the maximum horizontal orbital velocity at the crest of the reef.

We are interested in a maximum value of F which would occur at a certain value of θ. To perform this computation, we introduce simplifying parameters $\beta = (F_M/2F_D)$, $\alpha = u_o/u_m$, $\gamma = \sin \theta$, and $\lambda = \cos \theta$, and solve for:

$$\lambda_i(\gamma_i + \alpha) + \beta\gamma_i = 0 \qquad (13)$$

$$1 - 2\gamma_i^2 - \alpha\gamma_i + \beta\lambda_i < 0$$

The real roots for Eq. (13), γ and λ, which also satisfy the following Eq. (14), are substituted into Eq. (11) to obtain the maximum F value.

For most practical applications, one may assume a worst condition in which the current and wave actions agree in direction. In this case, we may introduce an apparent drag coefficient, C_D', such that

$$F = C_D' A w_o (u_o + u_m)^2/2g \qquad (14)$$

A nomogram such as in Figure 4.10 can be used to obtain C_D' from u_o and u_m.

1. Example

Using the same reef block as in the previous example (Fig. 4.8) with identical ambient conditions

Wave: Height 6.0 m, Period 12.0 sec
Current: Speed $= U_H = 0.5$ m/sec
Direction: 60° from wave direction
Water Depth: $h = 50$ m

The maximum horizontal velocity due to wave orbital motion at the crest of the reef (1.5 m above the seafloor) can be computed readily from procedures described by the U.S. Army Corps of Engineers (1984) as

$$u_m = 0.711 \text{ m/sec}$$

The current speed at the crest of the reef is

$$u_o = \cos \alpha + 1.5 u_H (z/h)^{1/7}$$
$$= 1.414 \times 0.5 \times (1.5/50)^{1/7}$$
$$= 0.43 \text{ m/sec}$$

Also,

$$F_D = 0.121 \text{ ton} \quad \text{and} \quad F_M = 0.071 \text{ ton}$$

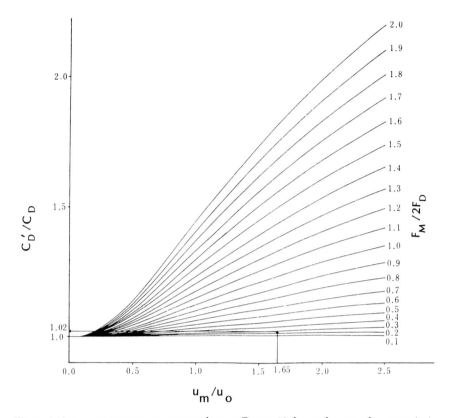

Figure 4.10 A nomogram giving apparent drag coefficient C_D' due to the coincident wave (u_m) and current (u_o) action. (Source: JCFPA, 1989.)

To obtain the maximum horizontal thrust from the nomogram in Figure 4.10, one computes the input parameters:

$$F_M/2F_D = 0.293 \quad \text{and} \quad u_m/u_o = 1.65$$

which leads to $C_D'/C_D = 1.02$, or $C_D' = 1.02 \times 2.0 = 2.04$. Thus, the maximum lateral thrust acting on the reef block due to the combined effect of wave and current is

$$
\begin{aligned}
F &= C_D' \, Aw_o(u_o + u_m)^2/2g \\
&= 2.04 \times 2.27 \times 1.03 \times (0.43 + 0.711)^2/2 \times 9.8 \\
&= 0.317 \text{ ton}
\end{aligned}
$$

The reef will slide horizontally when the lateral thrust of the fluid forces exceeds the resistance offered by the friction between the reef and the seafloor. To prevent sliding, the resistance must be greater than the horizontal thrust by the fluid, or

$$S_F = \sigma_G V\mu(1 - w_o/\sigma_G)/F \tag{15}$$

and

$$S_F \geq 1.2$$

in which S_F is the safety factor chosen to be greater than 1.2, and μ is the friction coefficient between the base of the reef block and the seafloor.

In order to prevent a toppling failure, the resisting moment must remain greater than the overturning moment, such that

$$S_F = \sigma_G V(1 - w_o/\sigma_G)s_w/(Fh_o) \geq 1.2 \tag{16}$$

in which h_o is the height of the lateral force F above the seafloor, and s_w is the distance between the point of rotation on the reef block and the center of gravity.

2. Another Example

Using the previous example in which the maximum lateral thrust caused by the combined effect of wave and current was $F = 0.317$ ton, one can evaluate safety against sliding from Eq. (15) as follows:

$$S_F = 2.44 > 1.2 \text{ (safe)}$$

And one can evaluate safety against toppling from Eq. (16) as follows:

$$S_F = 4.05 > 1.2 \text{ (safe)}$$

Note that in this example, values of $\mu = 0.6$, $s_w = 1.5/2$, and $h_o = 1.5/2$ were used.

D. Scour and Burial

Sediment motion is inevitable where the speed of water particles due to wave or current exceeds the threshold velocity of given sediment sizes. In an area subjected to active sediment motion, it is desirable for reef blocks to rest on three or four point contacts with the floor. Planar contact between the reef block and the seafloor is conducive to scour and eventually to toppling of the block. Where the seafloor consists of soft mud, the currents are generally too weak to cause significant scour. However, a structure may subside into the seafloor by compacting the loose mud by its own weight. In this case, the planar contact between the object and the seafloor is desirable (JCFPA, 1986).

VI. Biological Function and Design Criteria

The ideal approach to planning artificial reefs, of course, is to take advantage of the biological instincts of fishes—including their known behaviors in response to various foreign objects placed in the water. In our view, however, the accumulated knowledge of fish instincts and response to such objects is minimal.

Based on many years of experience and observations, Japanese scientists have become convinced that a fish's affinity to underwater objects is guided by its instincts, and that the degree of the affinity to the reef varies by fish species (Nakamura, 1985). In particular, they found that orientation or taxis of fish relative to various excitations is important in understanding the response of the fish to artificial reefs. For instance, rheotaxis guides the fish to orient itself parallel to the current; geotaxis guides the fish to balance its body relative to the bottom; thigmotaxis helps the fish to navigate relative to an object with which it is in physical contact; the fish also responds to light through phototaxis, and to smell through chemotaxis.

Affinity of fishes to reefs appears to vary widely depending upon the fish species and stages of maturity (Okamoto *et al.*, 1979a). With respect to artificial habitat, Nakamura (1985) recognized three categories of *Gyosho-Do*, a Japanese term that is translated in English as *reefiness* (Fig. 4.11) as follows: Type A fishes are benthic dwellers that prefer physical contact with the reef, occupying holes, crevices, and narrow spaces. Examples include greenling (*Hexagrammus otakii*), rockfish (*Sebastiscus marmoratus*), and brown rockfish (*Sebastes inermis*). Type B species surround the reef without direct physical contact with it, and they are believed to be linked to the reef through vision and sound. Examples include red seabream (*Chrysophrys major*), striped grunt (*Parapristipoma trilineatum*), brown rockfish (*Sebastes inermis*), flounder (*Paralichthys oblivaceus*), and dab (*Limanda herzensteini*). Type C species hover in the mid- and surface layers of the water at some distance from the reef. Examples include yellowtail (*Seriola quinqueradiata*), skipjack (*Katsuwonus pelamis*), dolphin (*Coryphaena hippurus*), and Pacific mackerel (*Pneumatophorus japonicus*).

A view held by some Japanese scientists is that because of poor vision, fishes cannot discern the geometrical relationship of objects more than about 2 m apart (Nakamura, 1985). This view is believed to have led to the practice, popular in Japan, of limiting spacings between structural members in a reef module to less than 2 m. Others (Y. Ogawa, personal communication, 1983) contend that vertical panels or walls, and to a lesser extent beams, offer greater attraction to fish than do poles or skeletal members or columns.

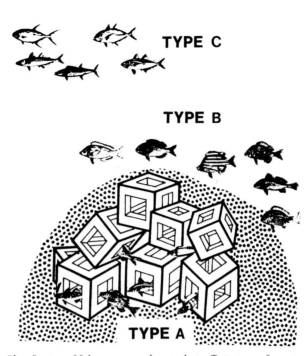

Figure 4.11 Classification of fish types according to their affinity to reef structure (from Nakamura, 1985).

They cite car bodies, often highly successful fish attractors, as evidence. Although much of our knowledge of fish habits relative to artificial habitats is essentially rule of thumb, it represents a synthesis of accumulated experience worthy of respect. At the least, it remains a useful working hypothesis.

In this volume, the ecology of artificial habitats is reviewed in Chapter 3, while their biological impacts are reported in Chapter 5, and practices of environmental assessment are described in Chapter 6.

A. Reef Design Based on Fish Behavior

Designing reefs may take advantage of the known affinity to reef habitat of fish species being targeted (Nakamura, 1985). For instance, for type A fishes the reef must provide internal spaces that will match typical sizes of the tenant fish. For type B fishes internal spaces of the reef must be large enough to allow visual recognition by the fish at a distance from the reef: about 1.5 m and up to 2 m in diam. Thus, in order to cater to both type A and B fishes, the internal spaces of the reef should vary in diameter with an upper limit of 2 m.

At night, fish tend to lose the ability to link themselves visually to the reef (Nakamura, 1980). Some fishes apparently manage to stay at the reef even in darkness by sensing pressure fluctuations in the water that are caused by vortex shedding from the reef in the presence of an ambient current. Occurrence of vortex shedding in a current with a velocity of u (cm/sec) requires the thickness (or width) of a structural member B (cm) to satisfy the following relationship (Nakamura, 1980):

$$Bu > 100 \text{ [cm}^2\text{/sec]} \tag{17}$$

For instance, for $u = 20$ cm/sec, B must be at least 5 cm.

The reef may be designed to keep the current speeds in the lee at acceptable levels for small fish. The current speed in the lee, u', is a function of the impinging current speed, u, and the structure of the reef, and may be approximated by:

$$u' = u(1 - C_D A/2S) \tag{18}$$

in which C_D equals a drag coefficient for the reef, A is the projection area of the physical portion of the reef, and S is the projection area of the maximum envelope of the reef. When the ambient current is strong, small type II category fish manage to stay close to the reef by taking refuge in its lee (Fig. 4.12).

When targeting type C fishes, the issue is whether or not the influence of the reef reaches the mid- and surface-layers of the water column. For instance, the turbulence generated by the reef may propagate upward in the water column, revealing the presence of the reef to type C fishes in the mid and upper layers. The upper limit to which turbulent fluctuations can rise beyond the crest of the reef has been measured with fishing sonar, as shown in Fig. 4.13. This shows that the turbulence rises rather rapidly to an elevation about 80% of the water depth as the reef height increases to about 10% of the water depth, but that further increases in reef height beyond 10% of the water depth bring diminishing increases in the upward excursion of turbulence. Figure 4.13 suggests that for type C fish, the adequate reef height is about 10% of the water depth (see JCFPA, 1986).

An example of a reef module that incorporates various experiences with fish habits is shown in Fig. 4.14. It has five identical side walls and a canopy with a hole in the center. The interior of the module is subdivided into five equal compartments with dividers, which also hide windows from each other. Impinging currents that pass through the windows would strike

Figure 4.12 Reduction in current speed in the lee of a submerged object. Legend: u: impinging current speed; S: projection area of the maximum envelope of the reef; u': current speed in the lee. (Source: JCFPA, 1986.)

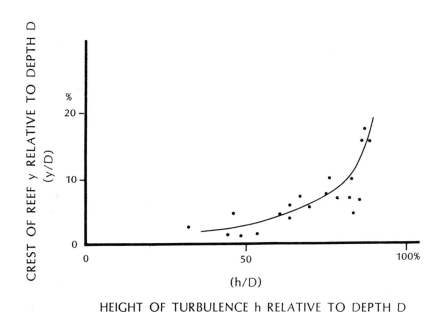

Figure 4.13 Relationship of the rise of turbulence in the water column versus reef height (from JCFPA, 1986).

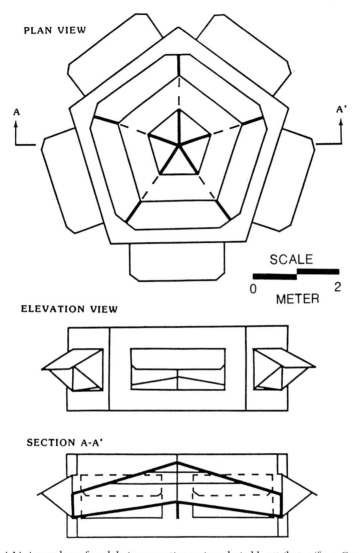

PLAN VIEW

A   A'

SCALE

0   2
METER

ELEVATION VIEW

SECTION A-A'

Figure 4.14 A sample reef module incorporating various desirable attributes (from Grove and Sonu, 1983).

the dividers and be deflected upward toward the hole in the canopy, creating local upwelling. The upward deflection of the current would also provide a measure of safety from scour off of the sea floor surrounding the module. The design allows for many surfaces, angles, and edges to encourage growth of algae.

Other interesting features of the module shown in Fig. 4.14 include:

Vertical panels. The exterior walls of the module and its interior dividers consist of vertical panels. In a hope to maximize visual stimuli to fishes, as much as 55% of all the material needed to construct this module is comprised of vertical panels, the remainder consisting of tilted panels (23%) and horizontal panels (22%).

Omni-directional shape. The pentagon-shaped plan view of the module gives it a virtual omni-directional response to impinging currents. Thus, the module need not be oriented into specific direction when installing. The pentagon shape is probably superior to a rectangle because of its rounded corners, and to a hexagon because of its simpler construction.

Shadows. Shadows from the sunlight are believed to accentuate the presence of a reef to the fish. The canopy as well as the side walls and compartment dividers of this module will cast shadows on the surrounding seafloor, if placed within shallow enough depths to allow sunlight penetration through the entire water column.

Enclosure of multiple spaces. The module contains subdivided interior spaces that can function either as semi-independent enclosures or connected chambers. This feature allows good maneuverability for the fish through and around the reef structures.

B. Planning Based on Hydrodynamic Principles

The lee wave was recognized early as an important factor whereby a reef located on the seafloor can make its presence felt in other parts of the water column (Fig. 4.15). Research on hydrodynamic aspects of artificial reefs has been most active in Japan (Sawaragi and Nochino, 1980; Sato, 1985; Nakamura, 1979, 1980, 1985; Imai et al., 1983; H. Sakuda et al., 1981; M. Sakuda et al., 1982; Matsumi and Seyama, 1985). The lee wave is a stationary, internal wave originating at an obstacle in the current in density-stratified water. Nakamura (1979) ascertained that a lee wave is most likely to develop when the densimetric Froude number, F_r, as defined below, is about 0.09 and the reef height is 10% of the water depth:

$$F_r = u^2/(wgh) \qquad (19)$$

Figure 4.15 Schematic for generation of lee wave due to a reef located in a density-stratified water column; h = height. (from Nakamura, 1985).

and

$$w = (\rho_2 - \rho_1)/\rho_2 \tag{20}$$

in which u is the current velocity, g the acceleration due to gravity, h the water depth, and ρ_1 and ρ_2 are the water density in the upper and lower layers, respectively. When high ambient current velocities force the densimetric Froude number, F_r, to exceed 0.32 ($= 1/\pi$), the lee wave is swept away and replaced by a current shadow such as the one shown in Fig. 4.16 (Sato, 1977; Wang and Sato, 1986; Wang et al., 1989). The current shadow tends to hug the bed, which probably explains (at least in part) why type C fishes are found in the lower part of the water column during a storm.

C. Fish Aggregation as a Function of Reef Design

A Japanese government-sponsored study aimed at quantifying fish aggregation around artificial reefs resulted in several reports (Higo, 1974; Higo and Nagashima, 1978; Higo and Tabata, 1979; Higo et al., 1979, 1980; Okamoto et al., 1981). Extensive underwater surveillance of fish aggregation was performed using diver inspection, diving research submarine, sonar probing, underwater photography, and both hand-held and remote-controlled video cameras (JCFPA, 1986). Partial results of this study reveal that bottom-dwelling fish species tend to exhibit aggregation within about 3 m

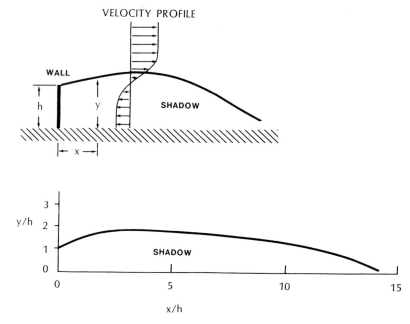

Figure 4.16 An example of current shadow behind an obstacle in a speedy current (from Sato, 1977).

above the sea floor, and even the midlayer fish tend to exhibit aggregation at only about 3.5 m from the reef. The findings suggest that the reef design should stress maximum horizontal spread of the reef, rather than the vertical dimension, and that artificial reefs probably need not be higher than approximately 5 m above the sea floor (JCFPA, 1986).

The research program also addressed the effective range of artificial reefs. For reefs with average bulk volume of about 3000 m³, the effective range of influence was about 200 m measured from the edge of the reef (JCFPA, 1986). The range of influence was considerably larger for bottom-dwelling species such as flounder, sole, and dab. Flounder were particularly sensitive to reefs; a tagged flounder was found to migrate readily between reefs as much as 900 m apart (JCFPA, 1986). For rock-dwelling species, maximum fish aggregation was accomplished with spacings between grouped reefs less than 400 m apart. Grouped reefs remained effective with spacings of as much as 1000 m (JCFPA, 1986).

The Japanese fish-aggregation study also revealed that fish aggregation, expressed in fish volume in kilograms per unit reef bulk volume (cubic meters), was about 5.0 for reefs consisting of concrete cylinders and hollow cubes, about 0.6 for large modular reefs, and as low as 0.03 for natural reefs (JCFPA, 1986). Reefs constructed with small concrete blocks appeared to support the highest diversity of species (JCFPA, 1986).

VII. Marine Ranching

The original concept of "marine ranching" dates back to 1977 when the Japanese government launched an ambitious new program called Coastal Fishing Grounds Development Program (*Engan Gyojo Seibi-Kaihatsu Jigyo*, usually abbreviated as *Ensei*). Its aim was to transform the Japanese fishery from the traditional practice of catching fish wherever they occurred, to one of nurturing and creating a harvestable fish stock in home waters. In a sharp departure from the formerly limited use of artificial habitat (which had primarily supplemented *existing* natural reefs) the new approach clearly recognized the enhanced role of artificial reef technology in creating new fishing grounds where none had previously existed.

Artificial reef technology has become an indispensible tool to the success of the future Japanese fishery referred to as *Tsukuru Gyogyo* meaning a fishery that *creates* (Fig. 4.17). The importance of artificial reefs in the Ensei program is evidenced by the fact that the budget for their installation in Japan increased nearly five-fold from 1976 (a year before the inauguration of the Ensei program) to 1987; i.e., from about U.S. $30 million a year to U.S. $135 million a year (Fisheries Agency, personal communication, 1989).

Marine ranching was conceived as the first major stepping stone toward the long-term goal of the *Tsukuru Gyogyo* under the new Ensei program. Marine ranching, officially called "a system capable of domesticating coastal fisheries resources," aims to maintain and manage the complete life cycle of high-valued fish stocks within a corral-like enclosure, similar to a cattle herd in a fenced ranch. A 9-yr program to develop methods to control and manage as well as understand the life cycles of selected species was initiated in 1980. Results have been summarized in a report by the AFFTC (1989). An initial three-year study (1980–1983) focused on how to improve survival rates of juvenile fishes in captivity, concentrating on such species as cherry salmon (*Oncorhynchus masou*), bluefin tuna (*Thunnus thynnus thynnus*), halibut (*Paralichthys olivaceus*), scallop (*Pecten albicans*), and bloody clam (*Scapharca broughtonii*). The second phase (1983–1986) was devoted to field testing results of the previous study, plus developing facilities to enhance and manage habitats for individual species. The third, 3-yr study (1986–1989) focused on grouping multiple species into a single mixed community

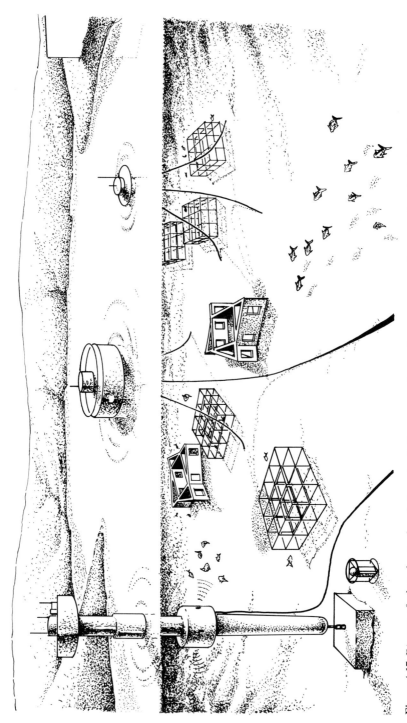

Figure 4.17 Diagram of selected water-based elements of a Japanese marine ranching system. Not shown are shore-based facilities.

compatible with a given geographical environment. A typical example of such a mixed community was a group of salmon, sea bream, abalone, and sea urchin.

To keep the domesticated species within a given geographical area, various techniques are being investigated. At Saiki Bay, on the southernmost island of Kyushu, Japan, where the first prototype marine ranch experiment was deployed in 1983, juvenile sea bream (*Chrysophrys major*) were trained to respond to acoustic meal calls (Grove *et al.*, 1989). Test results revealed that this method worked reasonably well for the first year after release of fish in the bay, as demonstrated by a recapture rate as high as about 10% for this period. However, the recapture rate declined sharply as the fish matured, to about 1% after the first year and to as low as about 0.1% after the second year.

It appears that the ability to barricade the fish in a given geographical area without a physical confine may be one of the key factors affecting the success of the marine ranching concept. Thus far, several ideas for a net-free fish "fence" have been suggested, such as flashing lights, laser beam, sound, electric current, and magnetic field. These concepts are investigated using prototype tests at two locations, Kumamoto and Kagoshima, on the island of Kyushu. A preliminary evaluation performed in a small-scale tank experiment (Kumamoto Prefecture, Fishery Development Department, 1990) indicates that the electric fence can accomplish virtually 100% efficiency in preventing fish from crossing the barrier. It was even suggested that fish are so sensitive to the electric charge that power may be turned on intermittently to save energy after the fish have been conditioned to the presence of the electric fence. In contrast, fish sensitivity to sound and light, including laser beam, tends to diminish over time.

VIII. Knowledge Gaps

From a historical perspective, discovery of the affinity of fishes for alien objects in the water initially led to the practice of installing artificial reefs to cause fish aggregation. After a long period of this primitive practice, in which various aspects of reef design, such as shape, size, type, material, grouping, placement pattern, and relation to existing natural reef were subject to experimentation. A new realization has evolved that to be truly successful in enhancing marine resources, one must attempt to create an optimum habitat for the fish, using artificial reefs only as a component in such a habitat. This new trend, as demonstrated in the recent Ensei program in Japan, calls for further improvement in the understanding of how artificial reefs function in the water; namely, why they attract fish, whether they enhance the biota in net terms or merely redistribute them, whether the reefs remain compatible with other attributes needed in the habitat, and above all, how the reef interacts hydrodynamically with the environment.

The cause of fish aggregation is important information that one day will make the design of a fishing reef a fully rational scientific process. Although various hypotheses have thus far been advanced regarding this subject, mainly related to factors such as behavior, feeding, and spawning, there has been very little focus on the hydrodynamic processes that would arise from the interaction between reef and the ambient environment, particularly the impinging current. For instance, according to a survey of eight fishing communities in the Tottori prefecture in Japan (Matsumi, 1988), 66% of those polled stated that fishes tend to aggregate on either the upcurrent and down-current side of the reef, and that they tend to alternate between these two sides as the current direction changes. About 20% of the same poll disagreed with this notion. The wake in the lee of a reef is expected to function as a trapping ground for drifting sea weeds, their spores, plankton (Kakimoto and Okubo, 1983), and juvenile fishes, but this role has not been sufficiently investigated scientifically. Future design may consider the importance of varied reef shapes and porosity to generate desired trapping characteristics in the lee of the reef.

Improved understanding of hydrodynamic characteristics of a reef will also benefit future design of reef deployment patterns. Reef clusters would interact with each other hydrodynamically, since wakes and other alterations (pressure fluctuations) to the currents caused by the reef propagate. This interaction is a function of spacing between reefs, relative positions, direction of impinging currents, current velocity and seafloor topography. Present knowledge of hydrodynamic interactions between adjacent reefs is inadequate to be incorporated in planning reef deployment.

Stability of reefs after placement on the seafloor is another area of insufficient knowledge. According to a survey of four fishing communities that have artificial reefs about 30 years old (Matsumi, 1988), 75% of those polled agreed that the reefs have gradually lost their function as an effective aggregation inducer over the years, and 46% in the same poll cited the subsidence or burial of reef as the possible reason. Scour around a reef and subsequent subsidence and burial have not received adequate investigation.

Excessive hydrodynamic loading caused by currents and waves on the reef may cause movement of a reef in place, and even if it does not lead to structural failure, a mere rocking motion, if repeated sufficiently, could facilitate scour and subsidence of the reef. Existing methods to estimate such hydrodynamic loading (i.e. those contained in the Japanese design manual (JCFPA, 1984)) are based on the well-known Morison formula (Morison et al., 1954). The drag coefficient and the virtual mass coefficient used in this formula assume a steady current impinging upon an object with simple shape. In the presence of an unsteady impinging current, such as the type associated with wave orbital motion, not only may these coefficients vary considerably, but also there will be an uplift force due to the vertical component in the wave orbital motion. The need to modify the coefficients also would arise from the fact that the artificial reefs can have complex configurations. Very little research has been performed on these subjects, and additional studies are essential to furthering the present knowledge base to guarantee optimal reef design.

References

Agricultural, Forestry, and Fishery Technology Council (AFFTC) 1989. Marine ranching. Koseisha-Koseikaku Publisher, Tokyo. (In Japanese)

Alevras, R. A., and S. J. Edwards. 1985. Use of reef-like structures to mitigate habitat loss in an estuarine environment. Bulletin of Marine Science 37:396.

Anonymous. 1989. Refuse-to-reefs: Old platforms process drilling wastes, garbage. Offshore (April 1989):42–44.

Bell, M., C. J. Moore, and S. W. Murphey. 1989. Utilization of manufactured reef structures in South Carolina's marine artificial reef program. Bulletin of Marine Science 44:818–830.

Binkowski, F. P., 1985. Utilization of artificial reefs in the inshore areas of Lake Michigan. Pp. 349–362 in F. M. D'Itri, editor. Artificial reefs: Marine and freshwater applications. Lewis Publishers, Inc., Chelsea, Michigan.

Bombace, G. 1989. Artificial reefs in the Mediterranean Sea. Bulletin of Marine Science 44: 1023–1032.

Buckley, R. M., D. G. Itano, and T. W. Buckley. 1989. Fish aggregation device (FAD) enhancement of offshore fisheries in American Somoa. Bulletin of Marine Science 44:942–949.

Burchmore, J. J., D. A. Pollard, J. D. Bell, M. J. Middelton, B. C. Pease, and J. Matthews. 1985. An ecological comparison of artificial and natural rocky reef fish communities in Botany Bay, New South Wales, Australia. Bulletin of Marine Science 37:70–85.

Chang, K. H. 1985. Review of artificial reefs in Taiwan: Emphasizing site selection and effectiveness. Bulletin of Marine Science 37:143–150.

Chang, K. H., and K. Shao. 1988. The sea farming projects in Taiwan. Acta Oceanographica Taiwanica 19:52–59.

Duffy, J. M. 1974. California's artificial reef experiences. Pp. 47–48 in L. C. Colunga and R. Stone, editors. Proceedings of an International Conference on Artificial Reefs, TAMU-SG-74-103. Texas A&M University Sea Grant College Program, College Station.

Fabi, G., L. Fiorentini, and S. Giannini. 1989. Experimental shellfish culture on an artificial reef in the Adriatic Sea. Bulletin of Marine Science 44:923–933.

Futch, C. R. 1981. An overview of state programs. Pp. 33–36 in D. Y. Aska, editor. Artificial reefs: Conference proceedings, Report 41. Florida Sea Grant College, Gainesville.

Grove, R. S., and C. J. Sonu. 1983. Review of Japanese fisheries reef technology, Report 83-RD-137. Southern California Edison Company, Rosemead, California.

Grove, R. S., and C. J. Sonu. 1985. Fishing reef planning in Japan. Pp. 189–251 in F. M. D'Itri, editor. Artificial reefs: Marine and freshwater applications. Lewis Publishers, Inc., Chelsea, Michigan.

Grove, R. S., C. J. Sonu, and M. Nakamura. 1989. Recent Japanese trends in fishing reef design and planning. Bulletin of Marine Science 44:984–996.

Helvey, M., and R. W. Smith. 1985. Influence of habitat structure on the fish assemblages associated with two cooling-water intake structures in Southern California. Bulletin of Marine Science 37:189–199.

Higo, N. 1974. On the fish attracting effect of artificial reefs ascertained by diving: I. Off the Katsuren Peninsula in Okinawa Prefecture. Memoirs of the Faculty of Fisheries, Kagoshima University 23:12–28. (In Japanese)

Higo, N., and M. Nagashima. 1978. On the fish gathering effect of the artificial reefs ascertained by the diving observation. 2. At the Sea Off the Satsuma Peninsula in Kagoshima Prefecture. Memoirs of the Faculty of Fisheries, Kagoshima University 27:117–130. (In Japanese)

Higo, N., and S. Tabata. 1979. On the fish gathering effect of the artificial reefs ascertained by the diving observation. 4. At the Off Sea in the West of the Biro Island in the Shibushi Bay. Memoirs of the Faculty of Fisheries, Kagoshima University 28:101–117. (In Japanese)

Higo, N., H. Hashi, S. Takata, and T. Kamimizutaru. 1979. On the fish gathering effect of the artificial reefs ascertained by the diving observation. 3. At the Off Sea of Taniyama, Kagoshima City. Memoirs of the Faculty of Fisheries, Kagoshima University 28:91–105. (In Japanese)

Higo, N., H. Hashi, I. Takahama, S. Tabata, M. Nagashima, S. Sakono, T. Kamimizutaru, and T. Yamasaki. 1980. On the fish gathering effect of the artificial reefs ascertained by the diving observation. VII. At the Off Sea of Makurazaki City. Memoirs of the Faculty of Fisheries, Kagoshima University 29:51–63. (In Japanese)

Imai, Y., O. Sato, K. Nahimoto, and K. Yamamoto. 1983. Fundamental studies on the scour around artificial reefs in a seabed. Bulletin of the Faculty of Fisheries, Hokkaido University 34:20–29. (In Japanese)

Ino, T. 1974. Historical review of artificial reef activities in Japan. Pp. 21–23 in L. C. Colunga and R. Stone, editors. Proceedings of an International Conference on Artificial Reefs, TAMU-SG-74-103. Texas A&M University Sea Grant College Program, College Station.

Japan Coastal Fisheries Promotion Association [Zenkoku Engan-Gyogyo Shinko-Kaihatsu Kyokai (in Japanese)] (JCFPA). 1982. Steel reef design and fabrication standards ("Kozai-Gyosho-yo Sozai Hyojun-Shiyo"). (In Japanese)

Japan Coastal Fisheries Promotion Association (JCFPA). 1984. Coastal Fisheries Development Program structural design guide ("Engan-Gyojo Seibi-Kaihatsu-Jigyo Kozobutsu Sekkei-Shishin"). (In Japanese)

Japan Coastal Fisheries Promotion Association [Zenkoku Engan-Gyogyo Shinko-Kaihatsu Kyokai (in Japanese)] (JCFPA). 1986. Artificial reef fishing grounds construction planning guide ("Jinko-Gyosho-Gyojo Zosei Keikaku-Shishin"). (In Japanese)

Japan Coastal Fisheries Promotion Association [Zenkoku Engan-Gyogyo Shinko-Kaihatsu Kyokai (in Japanese)] (JCFPA). 1989. Design examples for Coastal Fisheries Development Program structural design guides. ("Engan-Gyojo Seibi-Kaihatsu-Jigyo Kozobutsu-Sekkei Keisan-Rei-Shu"). 398 pp. (in Japanese)

Kagoshima Prefecture. 1987. Annual report on feasibility study for comprehensive development of nearshore and offshore regions. Kagoshima Prefecture, Japan. (In Japanese)

Kakimoto, H., and H. Okubo. 1983. Distributions of plankton around artificial reefs. Fisheries Engineering (Suisan Doboku) 19(2):21–28. (In Japanese)

Kumamoto Prefecture, Fishery Promotion Department. 1990. Proof-of-concept experiments on fishing fencing system. Aqua-Culture (Yoshoku) (February, 1990):81–83. (In Japanese)

Lewis, E. S., and K. K. McKee. 1989. A guide to the artificial reefs of southern California. State of California, Sacramento.

Matsumi, Y. 1987. Study on dynamic response of tethered submerged structures under wave action. Report of Coastal Engineering Laboratory, Osaka University. (In Japanese)

Matsumi, Y. 1988. Basic study on hydrodynamic functions of artificial reef. Report of Coastal Engineering Laboratory, Osaka University. (In Japanese)

Matsumi, Y., and A. Seyama. 1985. Flow characteristics around grouped artificial reefs. Proceedings of the 32nd Annual Japanese Coastal Engineering Conference: 652–656. (In Japanese)

Morison, J. R., J. W. Johnson, and M. P. O'Brien. 1954. Experimental studies on pile. Proceedings of the 4th Conference on Coastal Engineering: 340–370.

Mottet, M. G. 1981. Enhancement of the marine environment for fisheries and aquaculture in Japan, Technical Report 69. State of Washington, Department of Fisheries, Olympia.

Mottet, M. G. 1985. Enhancement of the marine environment for fisheries and aquaculture in Japan. Pp. 13–112 in F. M. D'Itri, editor. Artificial reefs: Marine and freshwater applications. Lewis Publishers, Inc., Chelsea, Michigan.

Nakamura, M. 1979. Civil engineering for fisheries ("Suisan Doboku Gagu"). INA Kogyo Tsushin-Sha, Tokyo. (In Japanese)

Nakamura, M., editor. 1980. Fisheries engineering handbook ("Suisan Doboku"). Fisheries Engineering Research Subcommittee ("Suisan Dopboku Kenkyu-Bukai"), Japan Society of Agricultural Engineering. Midori-Shobo Press, Tokyo. (In Japanese)

Nakamura, M. 1985. Evolution of artificial fishing reef concepts in Japan. Bulletin of Marine Science 37:271–278.

Nakamura, M., M. Uuekita, and T. Iino. 1975. Study on the landing impact on free-falling object in the ocean. Proceedings of the 22nd Annual Japanese Coastal Engineering Conference. (In Japanese)

Okamoto, T. Kuroki, and T. Murai. 1979a. Preliminary studies on the ecology of fishes near artificial reefs—Outline of the artificial reefs off Sarushima Island. Bulletin of the Japanese Society of Scientific Fisheries 45(6):709–713.

Okamoto, M., T. Kuroki, and T. Murai. 1979b. Fundamental studies on the ecology of fishes near artificial reefs: I. Preparatory observation of fish amount. Bulletin of the Japanese Society of Scientific Fisheries 45(9):1085–1090.

Okamoto, M., O. Sato, T. Kuroki, and T. Murai. 1981. The effect of divers on the behavior of fishes. Bulletin of the Japanese Society of Scientific Fisheries. Fisheries 47(12): 1567–1573.

Parker, J. H., P. M. J. Woodhead, H. R. Carleton, and I. W. Duedall. 1984. Coal waste artificial reef program, phase 4A, CS 2574, Project 1341-1. Electric Power Research Institute, Palo Alto, California.

Phillips, S. H. 1990. A guide to the construction of freshwater artificial reefs. Sport Fishing Institute, Washington, D.C.

Prince, E. D., O. E. Maughan, and P. Brouha. 1985. Summary and update of the Smith Mountain Lake artificial reef project. Pp. 401–430 in F. M. D'Itri, editor. Artificial reefs: Marine and freshwater applications. Lewis Publishers, Inc., Chelsea, Michigan.

Quigel, J. C., and W. L. Thornton. 1989. Rigs to reefs—a case history. Bulletin of Marine Science 44:799–806.

Relini, G., and O. Relini. 1989. Artificial reefs in the Liguarian Sea (Northwestern Mediterranean): Aims and results. Bulletin of Marine Science 37:143–150.

Rutecki, T. L., J. A. Dorr, III, and D. J. Jude. 1985. Preliminary analysis of colonization and succession of selected algae, invertebrates, and fish on two artificial reefs in inshore southeastern Lake Michigan. Pp. 457–489 *in* F. M. D'Itri, editor. Artificial reefs: Marine and freshwater applications. Lewis Publishers, Inc., Chelsea, Michigan.

Sakuda, H., M. Sakuda, K. Watanabe, and H. Onishi. 1981. Basic study on hydrodynamic characteristics of artificial reef models. Fishery Engineering (Suisan Doboku) 18:7–19. (In Japanese)

Sakuda, M., T. Kuroki, Y. Takagi, A. Kawaguchi, and T. Fukuda. 1982. Experimental study on hydrodynamic characteristics of large-scale artificial reefs. Pp. 53–55 *in* Proceedings of Conference, Nippon University, Faculty of Science. (In Japanese)

Sato, O. 1977. Selected topics on artificial reef. Coastal Oceanography Research Note 14: 88–100. (In Japanese)

Sato, O. 1985. Scientific rationales for fishing reef design. Bulletin of Marine Science 37: 329–335.

Sawaragi, T., and M. Nochino. 1980. Vortex simulation behind an artificial fish-house due to tidal current and wave-induced current. Pp. 935–944 *in* International Conference on Water Resources Development. Taipei, Taiwan, Republic of China.

Seaman, W., Jr., and D. Y. Aska. 1985. The Florida reef network: Strategies to enhance user benefits. Pp. 545–562 *in* F. M. D'Itri, editor. Artificial reefs: Marine and freshwater applications. Lewis Publishers, Inc., Chelsea, Michigan.

Sheehy, D. J. 1983. Evaluation of Japanese designed and American scrap material reefs, Research and Development Report 83-RD-607. Aquabio, Inc., Belleair Bluffs, Florida.

Sheehy, D. J., and S. F. Vik. 1989. Extending mitigation banking beyond wetlands. Pp. 1242–1253 in O. T. Magoon, H. Converse, D. Miner, L. T. Tobin, and D. Clark, editors. Coastal Zone 89: Proceedings of the Sixth Symposium on Coastal and Ocean Management. American Society of Civil Engineers, New York.

Sonu, C. J., and R. S. Grove. 1985. Typical Japanese reef modules. Bulletin of Marine Science 37:348–355.

Stone, R. B. 1985. History of artificial reef use in the United States. Pp. 3–11 *in* F. M. D'Itri, editor. Artificial reefs: Marine and freshwater applications. Lewis Publishers, Inc., Chelsea, Michigan.

U.S. Army Corps of Engineers. 1984. Shore Protection Manual, Vols. 1 and 2. Coastal Engineering Research Center, Waterways Experiment Station, Vicksburg, Mississippi.

U.S. National Research Council. 1988. Fisheries technology for developing countries. Report of an ad hoc Panel of the Board on Science and Technology for International Development. National Academy Press, Washington, D.C.

Wang, C. H., and O. Sato. 1986. Hydrodynamic characteristics in simplified components of artificial reef structure. Bulletin of the Faculty of Fisheries, Hokkaido University 37: 190–206. (In Japanese)

Wang, C. H., O. Sato, K. Nashimoto, and K. Yamamoto. 1989. Hydrodynamic characteristics of eddy behind two arranged components of an artificial reef structure. Bulletin of the Faculty of Fisheries, Hokkaido University 40:182–192. (In Japanese)

Woodhead, P. M. J., J. H. Parker, and I. W. Duedall. 1985. The use of byproducts from coal combustion for artificial reef construction. Pp. 265–292 *in* F. M. D'Itri, editor, Artificial reefs: Marine and freshwater applications. Lewis Publishers, Inc., Chelsea, Michigan.

5

Fisheries Applications and Biological Impacts of Artificial Habitats

J. J. POLOVINA

Southwest Fisheries Center, Honolulu Laboratory
National Marine Fisheries Service
Honolulu, Hawaii

This chapter examines many of the applications of artificial habitats in various aquatic environments. The particular focus is on their biological impacts with reference to fisheries and their actual or possible role in fishery management.

Grouping artificial habitats by their primary users is a useful way to categorize the applications of these structures for analysis. Four principal categories recognized here include artisanal fisheries; small-scale commercial fisheries; recreational fisheries and diving; the replacement of habitat lost from shoreside development (mitigation); and enhancement of habitat in marine reserves. Examples of the uses and impacts of artificial habitats for these categories are discussed in the first section of this chapter.

The second section focuses on the biological impacts of artificial habitats. Their much debated role in aggregating production and creating new production is presented in a broader context. It is proposed that artificial habitats may (1) redistribute exploitable biomass without increasing it or total stock size; (2) aggregate previously unexploited biomass and increase exploitable biomass; or (3) increase total biomass. The discussion of each provides examples. Throughout the text, attention is given to identifying the effective uses of artificial habitats in fishery management.

I. Applications

Some of the better documented fishery and environmental applications of artificial habitats are presented in this section. The intent is not to provide an exhaustive global review, but rather to provide illustrative data for representative situations. This augments the description and synopsis provided in Chapters 1 and 2.

A. Artisanal Fisheries

Artisanal fishermen have developed many artificial reef and fish-aggregating device (FAD) designs to create fishing grounds close to their villages. Such structures traditionally have been constructed with materials of opportunity, for example, sticks, poles, bamboo, or bundles of brush, but are also frequently made from concrete and scrap tires.

In the Philippines, a widely used artificial reef module is made from bamboo poles arranged in a tripod, weighted with stones, and covered with coconut palm fronds (Fig. 5.1). The units are usually placed in calm, shallow coastal water. When such artificial habitats are used in deeper water, a FAD may be attached to mark their location and to attract pelagic fishes. Large-scale deployments of reefs and FADs are used in regional development programs (Fig. 5.2).

Approximately 16,000 pyramid bamboo modules in clusters of 50 have been set along 40 km of Philippines coastline in the central Visayan Islands, and over 8000 bamboo modules have been deployed in the Samar Sea-Ticao Pass Project (Miclat, 1988). These artificial habitats are planned, constructed, deployed, and maintained by the village fishermen (Miclat, 1988). The reefs are conveniently located, and the value of the catches during the first year of deployment can exceed the cost of the reefs and their installation (Miclat, 1988). Catches include caesionids, mullids, lutjanids, serranids, siganids, lethrinids, haemulids, acanthurids, and apogonids (Miclat, 1988). The bamboo pyramids used in the Central Visayan Project had an estimated installed cost of U.S. $4.00/m^3 and annual harvests of 8 kg/m^3; therefore, if all of the fish caught were sold, the installed cost of the reef would be recouped in nine months (Bojos and Vande Vusse, 1988). Recently, artificial reefs made from concrete reinforced with bamboo have been used instead of bamboo, which only has a 4-yr life span (Bojos and Vande Vusse, 1988).

Although FADs are used by artisanal fishermen in many countries, the situation in the Philippines is unique. The approximately 3000 FADs owned and deployed by commercial tuna purse seiners are also used by artisanal fishermen who handline around the FADs for large tunas (yellowfin, *Thunnus albacares*) swimming too deep to be caught in purse seine nets (Aprieto,

Figure 5.1 A benthic artificial reef used in the Philippines (redrawn from Aprieto, 1988).

1988). Because they exploit different resources and appropriate accommodations exist between commercial and artisanal fishermen, fisheries conflicts between these two user groups are rare.

In Cuba and Mexico artisanal fishermen use artificial reefs to attract lobster (*Panulirus argus*) and to facilitate their capture. In Cuba, flat layers of mangrove branches are used to form shelters (about 2 m in length and 2 m in width) that are set in depths of 4–6 m and raised about 10–15 cm above the ocean bottom by cross branches (Fig. 5.3). Fishermen shake the shelters and net the escaping lobsters. In the Gulf of Batabano, Cuba, cooperatives use 120,000 lobster shelters and harvest about 7000 metric tons (t) of lobster (U.S. National Research Council, 1988). In Mexico, similar shelters have been used since the late 1960s; many are now made from ferroconcrete and corrugated roofing material.

Thailand's Department of Fisheries has used old tires and concrete cubes

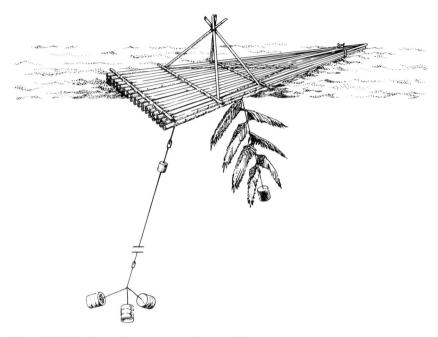

Figure 5.2 A bamboo raft or payao is a FAD used for tuna in the Philippines (redrawn from Aprieto, 1988).

Figure 5.3 A traditional Cuban lobster shelter (after U.S. National Research Council, 1988).

to construct artificial reefs for artisanal fishermen in the Gulf of Thailand. These reefs, placed on soft-bottom areas, provide hard substrata and vertical relief that attract valuable snappers (Lutjanidae) and groupers (Serranidae) normally not found at soft-bottom sites. In one application, the reefs were seeded with green mussels (U.S. National Research Council, 1988).

In another application, 2805 concrete cylinders were deployed in the Gulf of Thailand over a 41 km² area previously used by trawlers and village fishermen as a fishing ground for threadfin (*Eleuteronemus tetradactylum*, family Polynemidae). This artificial reef had the effect of closing the area to trawlers, thus allocating the resource to village fishermen using gill nets from small vessels (Sinanuwong, 1988). Before the deployment of this artificial habitat, village fishermen fished this resource for about 15 days in late November and early December, before the schools were depleted by trawlers and push-netters. However, after deployment, trawlers and push-netters were unable to operate in the area, and village fishermen were able to fish the schools for at least 6 months. The threadfin catch by village fishermen was 1746 kg (average catch rate, 4.7 kg/trip) before deployment and 5562 kg (average catch rate, 8.3 kg/trip) after deployment (Sinanuwong, 1988). Unfortunately, the substrata where the reefs were deployed were soft, and most of the reefs sank into the bottom sediment after about a year.

The Thai government is considering plans to increase its artificial reefs by using concrete cube modules (volumes of 1 and 2 m³) to construct large artificial reefs with volumes of 25,000–50,000 m³ and covering areas of 50–100 km² (Sungthong, 1988). These large reefs would close large areas to trawling and create fishing sites for artisanal fishermen.

By 1988, the Malaysian Department of Fisheries artificial habitat program had deployed 65 artificial reefs made from over 505,000 scrap tires, seven reefs made from sunken ships, and four reefs made from pyramids of concrete pipes (Hung, 1988). The tire reefs consist of modules of tires tied into pyramids with polyethylene rope. The number of tires per artificial reef site varied: most of the reefs (40%) had fewer than 1000 tires, but a few (5%) were composed of more than 30,000 tires (Hung, 1988). The objective is to enhance biological productivity and fishery resources in coastal waters. To prevent overfishing of resources aggregated at the artificial reefs, the Department of Fisheries prohibits fishing within 1.7 km of the artificial reefs (Hung, 1988). By 1990 the artificial reefs are expected to contain a total of two million tires.

B. Small-Vessel Commercial Fisheries

Small-vessel commercial fishermen typically use larger vessels with greater fishing power, hydraulics, depth finders, and inboard engines, than do artisanal fishermen. In many developed countries, they operate at marginal economic levels, and governments perceive programs of construction and deployment of artificial habitats as being beneficial to the financial operations of these fisheries.

Japan has the most extensive system of artificial reefs to assist the small-vessel commercial fishermen of any nation, as described in detail in Chapter 2. Since 1976, the Japanese government has spent over U.S. $1 billion for reefs with an enclosed volume exceeding 17 million m³ (Grove *et al.*, 1989), so that 9.3% of total nearshore seafloor to a depth of 200 m is covered with artificial reefs (Yamane, 1989). (See numerous illustrations in Chapter 4.)

The Japanese artificial reefs reportedly are popular with fishermen because they increase catches for a wide range of both demersal and pelagic fishes and decrease operating costs (Yamane, 1989). Whether the reefs actually increase fishery catches has been addressed by Polovina and Sakai (1989), who analyzed the 1945 to 1985 catch at a small bay in Hokkaido, Japan, where 50,000 m³ of artificial reefs were deployed from 1960 to 1985. They found that although several resources were caught, an increase in landings for only one resource could be attributed to the artificial reefs: catches of octopus (*Octopus dofleini*) increased by an estimated 1.8 kg/m³ of artificial reef. Fifty-three percent of the fishermen surveyed from this bay used the artificial reefs regularly, 12% used them only when fishing elsewhere was poor, and 36% did not use them at all (Polovina and Sakai, 1989). Further, 33% of these fishermen thought the reefs had expanded the amount of productive habitat, 38% thought the reefs did not increase the productive habitat, and 30% were unable to decide about this question.

In the Mediterranean Sea, Italy, France, and Spain have modest artificial reef and FAD development projects. The coastal environment in many parts of the Mediterranean Sea has a soft bottom, water with a high nutrient level that is not fully recycled by the ecosystem, and many nearshore fisheries that are overfished, in part, because of illegal trawling. The objectives of the projects include (1) protection of nursery grounds from illegal trawling, (2) attraction of pelagic and benthic species that use hard substrata, and (3) provision of substrata for shellfish farming and nutrient recycling in eutrophic environments. Initially, various materials including car bodies and ships were used as artificial reefs, but most recent and planned artificial reefs consist of concrete cubes or blocks (see Figs. 2.2 and 2.3).

In Italy, Bombace (1989) found that a concrete reef of 4300 m³ increased both mussel and fish catches including striped mullet (*Mullus barbatus*), meagre (*Arnyrosoma regius*), sea bass (*Dicentrarchus labrax*), and mussels (*Mytilus galloprovincialis*). The net proceeds for a fisherman operating within the reef were 2.5 times greater than those operating outside the reef. In eutrophic waters such as those of the Adriatic, the cost of the reefs was recovered about three times in seven years (Bombace, 1989). Initially, trawlers were opposed to the reefs because the area was closed to trawling, but their attitude changed as their catches increased along the edges of the reef zone (Bombace, 1989).

When artificial reefs are deployed in Italy, the area covered by reefs typically is designated as a marine zone, and activities and users in it are regulated. However, the demands to harvest resources in these zones often exceed their productivity, and administrators are faced with the challenge of allocating resources among users (Bombace, 1989). Commercial fishermen in Italy are promoting the development of more marine zones protected by artificial reefs; France has less interest in developing new zones, and Spain is just beginning to evaluate artificial reefs (Bombace, 1989).

Insular nations also have deployed artificial reefs and FADs to assist commercial fishermen. For example, 19 areas around Taiwan have artificial reefs built from concrete blocks deployed in 20–40 m depths on flat, sandy, or pebble bottoms to improve fishing sites (Chang, 1985). In Jamaica, artificial reefs made from scrap tires weighted with rocks or concrete are used to create fishing grounds near fishing villages and to provide habitat in areas closed to fishing to protect spawning stock from overfishing (Haughton and Aiken, 1989).

Most South Pacific island governments use FADs widely to enhance catches of offshore pelagic fishes (Fig. 5.4). Evaluation of FADs in American Samoa showed that their use could significantly increase catch per unit effort (CPUE) of offshore pelagic fishes for a troll fishery (Buckley *et al.*, 1989). However, replacing lost FADs is a permanent task for fishery departments since the life span of FADs anchored in unprotected ocean around Pacific islands is often only a few years. Further, FADs do not always increase catches significantly; a study in Puerto Rico found only a slight increase in catches with FADs (Feigenbaum *et al.*, 1989).

Increased CPUE due to artificial reefs and FADs alone may not justify their use by commercial fishermen when they receive heavy and unregulated usage. Since artificial reefs and FADs are usually located in accessible sites, they produce increases in fishing effort, possibly increase the catchability of the gear, and hence increase fishing mortality. Concern has been expressed that even for pelagics, overfishing may occur as a result of the increase in fishing mortality (catch) arising from the use of FADs (Floyd and Pauly, 1984). However, even if artificial reefs and FADs do not have a detrimental impact on the exploited stocks, they still may not be beneficial economically. An economic study of commercial open-access fisheries around FADs in Hawaii found that even if high levels of fishing at FADs do not result in recruitment overfishing, and if fishing effort is unregulated, installation of FAD networks will not generally increase fishermen's aggregate

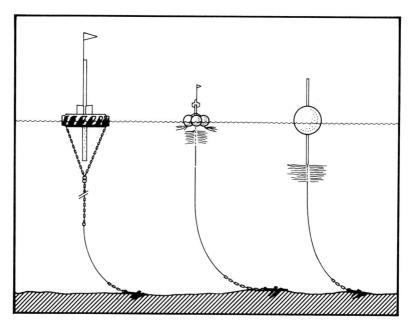

Figure 5.4 Representative FADs used in the Pacific Ocean (redrawn from U.S. National Research Council, 1988).

profit (Samples and Sproul, 1985). Further, deployment of FADs could result in decreases in employment, harvest levels, and sustained gross revenues. Limiting the commercial fishing effort at FADs is seen as a means of preventing these detrimental impacts (Samples and Sproul, 1985).

C. Recreational Fishing and Diving

Artificial reefs and FADs also are popular with recreational fishermen and divers because they provide convenient sites with a concentration of fishes and other organisms. Although globally they are not as widespread as artisanal and commercial fishing applications, where artificial habitats are employed recreationally, use can be extremely intensive. They often are constructed and deployed by sport fishing and diving organizations and state fishery departments in freshwater and marine settings. The most common materials used are ships, concrete, tires, and stone rubble (McGurrin *et al.*, 1989).

The most widespread recreational usage is in the United States. In the Gulf of Mexico, for example, 4000 petroleum structures function as artificial reefs (McGurrin *et al.*, 1989). Even while these structures are producing gas and oil, they are heavily used by recreational fishermen and SCUBA divers (Reggio, 1989). Louisiana has 3100 petroleum structures, which are the destinations of about 37% of all saltwater recreational trips, and over 70% of all recreational trips more than three miles offshore (Stanley and Wilson, 1989). A survey in southern Florida found that about 28% of the recreational fishermen and 14% of the sport divers regularly used artificial reef sites (Milon, 1989). Brush, timbers, tires, rocks, and concrete materials are used in lakes and reservoirs to enhance fishing (D'Itri, 1985). A large artificial reef covering 9500 m² was constructed in Smith Mountain Lake, Virginia, with 7000 scrap tires and 400 Christmas trees (Prince *et al.*, 1985). Sunfishes (*Lepomis* spp.) and white catfish (*Ictalurus catus*) were more abundant at this artificial reef site after deployment of reef materials. Furthermore, fishes foraged on the artificial reefs and catfishes deposited eggs inside the artificial reef (Prince *et al.*, 1985). A more detailed review of U.S. sport fishing habitats is provided in Chapter 2.

State government agencies in Australia also support artificial reefs for recreational fishing and diving. One structure is made of tires assembled in a tetrahedron to create fishing and diving sites, which are closed to professional fishermen (Young, 1988). In one instance, 34,000 tires, at a cost of A $205,000, were deployed on the premise that the artificial habitat would increase revenues in the local community through increased spending by sport fishing and diving interests (Young, 1988).

The reefs and FADs concentrate both fish and fishermen. Some concerns over the resource and the conflicts between users have been raised by Samples (1989) and others. Two forms of conflicts, i.e., competition over a common stock and conflicts from user congestion, have been observed (Samples, 1989). An example of the former occurs between commercial pole-and-line boats and recreational trollers around FADs in Hawaii. A pole-and-line vessel can capture all of the skipjack tuna (*Katsuwonus pelamis*) around a FAD, leaving nothing for recreational trollers in the short term.

Also, conflicts due to user congestion occur when many users are concentrated around a reef or FAD, often with various types of gear, such as purse seiners and trollers or handlining and diving gear. A number of approaches that restrict access, limit effort, or segregate users in space and time may resolve these conflicts (Samples, 1989). Of course, carefully planned artificial reefs and FADs also can serve to shift effort away from heavily used natural sites. Broader aspects of fishery management are discussed more fully in Chapter 7.

Artificial reefs for recreational uses have been constructed and deployed by fishing and diving clubs, which, unfortunately, may lack the resources or inclination to properly research the siting, design, and materials. Experience in Florida and other states indicates that the structures may be ineffective and even damaging (Andree, 1988). Andree (1988) has recommended that a Florida artificial reef plan be developed to establish standards for siting, design, and materials, and to establish central artificial reef permitting, maintenance, and monitoring systems. As other areas of the world initiate such programs, the experiences in active artificial habitat sites need to be consulted to avoid mistakes and negative environmental impacts.

D. Environmental Mitigation and Enhancement

Applications of artificial reefs (for uses other than increasing fishing success) include providing habitat to mitigate its loss due to coastal development or pollution and to improve habitat in marine reserves. This is a relatively recent application of this field, and much of the experience is limited to the Pacific mainland coast of the United States. In southern California, for example, the San Onofre nuclear station has affected organisms in two ways: by killing larval, juvenile, and adult fishes that are taken into the plant with the cooling water, and by producing a turbid plume that affects kelp, fishes, and invertebrates in the San Onofre kelp bed. A 120 ha artificial reef has been proposed as in-kind mitigation for impacts to the kelp-forest community from the plume, and a 60 ha structure proposed as out-of-kind mitigation for egg, larvae, and juvenile fish mortality from entrainment (Ambrose, 1990).

Loss of rocky habitat due to near shore filling was successfully mitigated with a 2.83 ha quarry rock artificial reef in Puget Sound, Washington (Hueckel et al., 1989). An important feature of this mitigation was that before a site for the habitat was selected, a set of benthic species was identified to predict colonization of the site by economically important fish species.

On the Pacific coast of Costa Rica, 5000 scrap tires were used to construct new habitat to protect marine fauna rather than for fishing (Campos and Gamboa, 1989). The reef, used by juvenile and adult fishes, was not marked, apparently to prevent fishermen from finding and fishing the area. Artificial reefs may have potential to protect or improve aquatic ecosystems. For example, in Maryland's Chesapeake Bay artificial reefs have been proposed to provide habitat for the American oyster (Crassostrea virginica) to restore oyster population levels, not for fishery harvest, but to filter excessive nutrient and particulate levels from the water (Myatt and Myatt, 1990).

In the case of mitigation, it is important to be sure that the artificial reefs are an appropriate habitat, are properly sited to replace the lost habitat, and are not adversely impacting other species. For example, species that use flat, or low-relief habitat may be adversely impacted if high-profile materials are deployed.

II. Biological Impacts

A considerable body of literature deals with the ecology at artificial habitat sites (see Chapter 3), whereas few studies address the biological impacts of artificial reefs and FADs on fish populations (Bohnsack and Sutherland, 1985; Bohnsack, 1989). The limited number of studies of this latter aspect is certainly not due to a lack of interest but rather to the difficulty in collecting the appropriate data. Data must be collected from large-scale applications of artificial habitat on an appropriate spatio-temporal scale to determine the possible biological impacts of artificial habitat in the presence of variations in the environment, fishing strategies, and gear. The scale of most research or pilot applications of artificial habitat is too small to detect biological impacts on stocks, even at a local level. Thus, much of the current thinking on the biological impacts of artificial habitat tends to be speculative. (The reader may consult Chapter 6 for a review of ecological assessment methods.)

Often discussions on the impacts of artificial reefs and FADs distinguish between impacts due to aggregation and those due to "new production" (Bohnsack, 1989). However, from both a management and biological perspective, it is important to make the distinction between aggregation that simply redistributes exploitable biomass and aggregation that attracts biomass not previously exploited, while increasing the exploitable biomass. It is useful to consider the three types of impacts on the exploitable biomass and the total stock due to the artificial habitat:

- Artificial reefs and FADs can simply redistribute the exploitable biomass without increasing it or total stock size;
- They can aggregate previously unexploited biomass and increase the exploitable biomass but not the total stock size;
- When stocks are limited by high-relief habitat, artificial reefs can increase stock and hence, total stock size.

These three types of impacts, therefore, distinguish between not only the two types of aggregation but also total stock size and exploitable biomass. Because some of their biological aspects will differ, the three impacts are discussed separately in subsequent sections. However, all three may occur in varying degrees in any artificial habitat application. They are illustrated in Fig. 5.5.

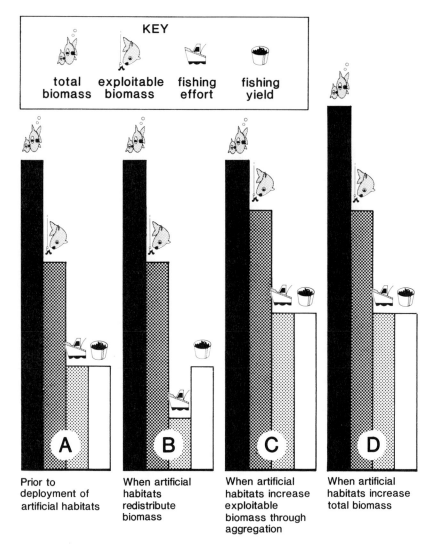

Figure 5.5 Three possible impacts of artificial habitat. (A) Total biomass, exploitable biomass, fishing effort, and yield for a resource prior to deployment of artificial habitats. Note the catch rate (yield:effort) is 1.0, and the yield is about one third of the total biomass. (B) When artificial habitats just redistribute the exploitable biomass to make it easier to catch, the same catch can be obtained with lower effort. (C) When artificial habitats increase the exploitable biomass but not the total biomass, an increase in catch can be achieved with greater effort without a reduction in catch rate, assuming that recruitment overfishing does not occur. (D) When artificial habitats increase the total biomass, the levels of all the variables in part (A) increase. Note that the only difference between (C) and (D) is the increase in total biomass.

A. Impacts Due to a Redistribution
of Exploitable Biomass

For some resources, artificial habitat may primarily change the distribution of the exploitable biomass without increasing it or the total stock (Fig. 5.5 part B). For example, some of the resources exploited in the natural habitat move to the artificial habitat, or a highly mobile resource that moves between natural habitats may visit artificial habitats as well.

This type of impact appears to be illustrated by flatfishes (Pleuronectidae) in the study by Polovina and Sakai (1989) on the impacts of 50,000 m³ of artificial reefs deployed in Shimamaki Bay off Hokkaido, Japan. Despite the flatfishes representing an estimated 30% of the gill-net catches at the artificial reefs, no increase in flatfish landings could be attributed to the reefs when landings from the entire bay were considered (Polovina and Sakai, 1989). In sonic tagging experiments, flatfishes readily moved from natural habitat to artificial reefs, but they were not long-term residents at either site (Kakimoto, 1984). Polovina and Sakai (1989) concluded that the artificial reefs redistributed the flatfishes but did not change their exploitable biomass.

Although the Shimamaki study did not identify any biological impacts from the redistribution of exploitable biomass, there are potential impacts caused by artificial reefs that redistribute exploitable biomass. The greatest potential impact may be a reduction in exploitable biomass if fishing at the artificial habitat is not restricted. Siting of artificial habitats usually allows them to be more accessible to fishermen all year and often works to support a higher density of fishes than a natural habitat. The higher density may increase catchability of the fishing gear, and the greater accessibility increases fishing efforts, which can result in higher fishing mortality. An increase in fishing mortality will decrease exploitable biomass in the area. Whether this decrease results in lower catches or recruitment to the fishery, either locally or in an adjacent region, depends on the stock dynamics. If the stock is migratory, then heavy fishing mortality in one region will result in lower levels of exploitable biomass in adjacent regions. If a strong regional stock–recruitment relationship exists, then heavy local fishing mortality could reduce future recruitment.

Similar to the situation in Shimamaki Bay, application of artificial reefs in the Gulf of Thailand also did not appear to increase the exploitable biomass for one resource. However, the allocation of the resource among user groups was altered (Sinanuwong, 1988). The reef site in the Gulf of Thailand was closed to trawlers and push-netters, and the threadfin resource was allocated to small-vessel, village fishermen using gill nets (Sinanuwong, 1988). Village fishermen had previously fished this resource for a short time until the schools were depleted by trawlers and push-netters. After the reef deployment, they were able to fish the schools much longer and catch, as well as catch rate, increased.

The Thailand example illustrates an application of artificial habitat that likely resulted in a reduction in fishing mortality, because the increase in catches by the village fishermen was probably less than the catches previously taken by the more efficient trawlers and push-netters. This example also shows that artificial habitat can result in a change in the types of fishing gear used. Such a change may impact the species caught, catchability, and fishing mortality. Since the species composition of the catches at the artificial habitat may differ from that at the natural habitat, fishing mortality may increase for some species but decrease for others as effort shifts from natural to artificial habitat.

B. Impacts Due to Increased Exploitable Biomass but Not to Total Stock Size

Aggregation may not only cause a resource to be redistributed but may also increase the biomass of a resource exploited by a fishery. (See Fig. 5.5 part C.) If artificial habitat aggregates juveniles, thereby making them more accessible to capture, the exploitable biomass may increase as the size of the fish at entry to the fishery decreases. Conversely, aggregation may make available to fishing gear a portion of the resource that has been distributed at a low density and has not been previously exploited. An extreme case would be a resource that has not been fished because it is widely distributed at a low density at a natural habitat. Artificial habitat will aggregate the resource at a density sufficient to support a fishery, and the resource can then be exploited. From a fisherman's perspective, if a resource is not overexploited, it does not matter whether exploitable biomass is increased by aggregating unexploited biomass to artificial habitat or from new production that increases stock size. In both cases, increased catches will be achieved without increased effort.

The impact of FADs on tuna in the Philippines appears to represent this type of aggregation. There, devices known as payaos (Fig. 5.2), together with purse seines or ring nets, were introduced to the tuna fishery in the early 1970s. As a result, skipjack tuna (*Katsuwonus pelamis*) catches rose from less than 10,000 t in 1970 to 266,211 t in 1986, representing 20% of the national marine catch (Aprieto, 1988). Over 90% of the tuna caught at the FADs were less than one year old, and they were about one-half the length of a mature tuna (Aprieto, 1988). There was some concern that the heavy fishing mortality with small length at entry may result in growth overfishing (i.e., lower catches than could be achieved with larger length at entry or lower fish mortality) and recruitment overfishing (i.e., a decline in recruitment to the fishery) (Aprieto, 1988).

The question of whether FADs can cause growth overfishing has been examined by Floyd and Pauly (1984). Four factors are necessary for growth overfishing: (1) presence of small fish on the fishing ground; (2) use of gear capable of catching small fish; (3) a market for small fish; and (4) high exploitation rates. All four of these factors are present in the tuna fishery in the Philippines. Using an exploitation rate of 0.7–0.8 for skipjack tuna and yellowfin tuna (*Thunnus albacares*) with the Beverton and Holt (1966) yield equation, the yield per recruit declines by an estimated 50% when the size at entry drops from one-half to one-fourth the asymptotic length (Floyd and Pauly, 1984). The recent decline in landings from this fishery may be partly due to growth overfishing (Floyd and Pauly, 1984). Further, analysis of stomach contents suggests that the predation on juvenile tunas by adult tunas is greater at FADs than in schools in the open ocean, suggesting that FADs can increase natural mortality as well (Aprieto, 1988).

The biological impacts of this type of aggregation include all of the impacts associated with aggregation that simply redistributes exploitable biomass. However, when aggregation increases the exploitable biomass, other impacts depend on the dynamics between aggregated and unaggregated fish. Clark and Mangel (1979) developed a model for tuna purse seining in which tunas move from subsurface populations to surface schools that are fished. Applying this model to the tuna fishery with FADs shows that the potential impact of FADs on the stock depends primarily on whether the rate of movement from the unaggregated population to the FADs, as well as the mortality of the unaggregated population, exceeds the intrinsic rate of population increase (Samples and Sproul, 1985). If the rate that tunas aggregate at FADs plus non-FAD mortality exceeds the population growth rate, then high fishing mortality at the FADs alone can drive the fishable population to zero. The relationship between catches at a FAD and effort follows the typical dome-shaped production curve. The biological impact is that excessive fishing effort at FADs can result in recruitment overfishing. However, when the growth of the population exceeds the non-FAD mortality and the rate of aggregation to the FADs, no amount of fishing at the FADs can exhaust the total population (Samples and Sproul, 1985). In this case, catch increases with effort to an asymptotic value, and the biological impact is that increasing fishing effort on aggregations cannot increase the fishing mortality beyond a certain level. While the Clark and Mangel (1979) model has been applied specifically to the tuna fishery at FADs, the results also apply to demersal resources aggregated at artificial reefs.

C. Impacts Due to Increased Total Stock Size

In theory, providing additional habitat could increase the population size for some habitat-limited stocks (Fig. 5.5 part D). For example, the habitat provided by artificial reefs might result in substrata for additional food, shelter from predation, settlement habitat, and lower densities at natural reefs (Bohnsack, 1989). However, despite the large number of studies on artificial reefs, very little direct evidence indicates that artificial reefs can increase the population size of a fish stock (Bohnsack, 1989).

One unplanned experiment in the United States that may merit further study is the biological impact from the oceanic petroleum platforms off Louisiana. A single petroleum platform in a depth of 40–60 m can provide about 1 ha of hard substrate, and platforms are estimated to represent over 90% of all hard-bottom substrate off Louisiana (Scarborough-Bull, 1989). The ecosystems at these structures differ from the naturally occurring soft-bottom ecosystem and demonstrate that artificial reefs can result in the establishment of hard-substrate ecosystems, even when isolated from similar ecosystems (Scarborough-Bull, 1989). Since these platforms represent large-scale habitat alteration with apparent impacts on species composition and abundance, plus fishing areas and species targeted, a quantification of these impacts would greatly add to our understanding of the impacts of artificial habitats.

In Japan, a relatively large-scale application of artificial reefs provides some evidence that artificial reefs can increase the total stock. A significant increase in landings and catch rates of *Octopus dofleini* was observed in a small bay near Shimamaki, Hokkaido, Japan after almost 50,000 m³ of artificial reefs were deployed (Polovina and Sakai, 1989). Additionally, availability of data from two adjacent regions in the same bay made it possible to compare relative changes in catches and catch rates as a function of artificial reef volume in each bay. While changes in environment or fishery economics could alter catches and catch rates in each region, the relative catches and catch rates should be unaltered by these factors and reflect only the impacts due to the artificial reefs.

The magnitude of the increase in octopus catches attributed to the artificial reefs was about 90 t or about 1.8 kg/m³ of artificial reef per year. Polovina and Sakai (1989) concluded that the artificial reefs increased the exploitable biomass of octopus. This increase may have come from either an aggregation of octopus from habitat not previously exploited or from new biomass due to the additional habitat.

Unfortunately, no surveys of octopus abundance and their size structure (over the natural habitat and artificial reefs before and after the deployment) were conducted to complement the fishery data and determine whether the reefs were aggregating the octopus or actually increasing the population size. However, Polovina and Sakai (1989) addressed this issue by examining the change in catches in the two adjacent regions. They hypothesized that, if artificial reefs aggregated octopus from the entire bay, then as the octopus moved to the region with the large reef volume, an increase in catches in the region with the large volume of artificial reefs would be accompanied by a corresponding decline in catches in the adjacent region with the low volume of artificial reefs. But if the artificial reefs increased the population of octopus, changes in catches in each region would be independent of the artificial reef volume in the adjacent region and depend only on the volume within each region. The catch and effort data indicated that the catches in each region were independent of the artificial reef volume in the adjacent regions, consistent with the hypothesis that the artificial reefs did indeed increase the population of octopus (Polovina and Sakai, 1989).

Studies on the ecology of *O. dofleini* have found that the animals are almost always associated with dens, with one animal per den. So in areas without a sufficient number of dens, habitat could be limited (Hartwick *et al.*, 1978).

When the exploitable biomass in a region is heavily fished, the density of the resource above the size at entry to the fishery is very low relative to the preexploitation density. Thus, habitat is not likely to limit the population above the size at entry to the fishery, and artificial reefs that provide more habitat for this portion of the population are not likely to increase new production.

If artificial reefs are to increase new production of this resource, they might provide habitat to improve larval settlement, juvenile growth, and a reduction in juvenile natural mortality. Thus, biological impacts of artificial reefs that increase total stock size are likely to include one or more of the following: an increase in postlarval settlement, juvenile growth, and juvenile survival. However, just as with the impacts from aggregation, an increase in fishing effort, and hence, fishing mortality may also occur as the fishery responds to more accessible habitat and higher catches.

1. Estimation of Biomass Increase

In the absence of studies that quantify an increase in stock size due to artificial reefs, two simple approaches—one based on yield from natural habitat and the other based on the standing stock estimates at artificial reefs, together with an estimate of yield to biomass—can provide useful estimates of the maximum potential enhancement due to artificial reefs.

For the first approach, yield per area of artificial reefs is simply estimated from fishery yield per area of corresponding natural habitat; the resulting figure is adjusted upwards for the observed higher catches between artificial reefs and natural habitat. For example, to estimate potential fishery catches from artificial reefs in the tropics, the range of fishery production from coral reefs must first be considered. Annual fishery production per area of coral reef habitat ranges from <1 t/km^2 to 18 t/km^2, with values clustering around 5 t/km^2 (Marten and Polovina, 1982). Biomass on artificial reefs in several tropical and subtropical studies is, on average, seven times greater than on natural habitat (Stone et al., 1979). If, for example, artificial reefs can support 10 times the exploitable biomass of natural coral reefs, then an average annual value for the fishery catches from an artificial reef in the tropics is 50 t/km^2 or 10 times those from a coral reef. This value is equivalent to a yield of 0.05 kg/m^2. If, as an upper bound, this yield is assumed to come from only 1 m of vertical relief, then the yield per artificial reef volume is 0.05 kg/m^3.

The second approach to estimating the new production of an artificial habitat uses the biomass estimated from local artificial reefs and then estimates the potential fishery yield as a fraction of that biomass. The Beverton and Holt (1966) yield equation can be used to determine the fraction of the biomass at the reefs that can be harvested on a sustainable basis, if estimates of a number of population parameters are available (Beddington and Cooke, 1983). However, in the absence of estimates of population parameters, an upper bound for sustainable catch can be taken as one-half the product of natural mortality and unexploited exploitable biomass ($0.5 \cdot M \cdot B_0$), where M is the natural mortality and B_0 is the unexploited exploitable biomass (Beddington and Cooke, 1983).

For example, the range of biomass estimates observed for tropical and subtropical artificial reefs is 26–698 g/m^2 (Stone et al., 1979). More recently, a value of 1266 g/m^2 was documented (Brock and Norris, 1989). Taking an average value for this range of 650 g/m^2 as an average estimate of the unexploited exploitable biomass, the fishery catches can be estimated by multiplying this value by an estimate of $0.5 \cdot M$. For a tropical, fast-growing, short-lived species, M equaling 0.7 might be appropriate. A biomass of 650 g/m^2 at the artificial reefs would then support a maximum annual fishery production of about 35% of the unexploited exploitable biomass, or 228 g/m^2. Again, if yield per square meter is assumed to be due to just 1 m of vertical relief, then in this example, 0.2 kg/m^3 is the upper bound for the potential fishery yield from artificial reefs.

Once an estimate of the fishery production due to new production from the artificial reefs is available for a specific application, this estimate can be compared with the actual catches from the reefs to determine to what extent they are functioning as fish aggregators. Total catches at artificial reefs have been documented at 8 kg/m^3 of artificial reef from the Philippines and 5–20 kg/m^3 from Japan (Sato, 1985; Bojos and Van de Vusse, 1988). Of course these catches include fishes aggregrated by the reefs as well as any new production due to the reefs. If the range of catches of 5–20 kg/m³ of artificial reef represents a range for tropical applications, then based on the example previously considered, estimates of new production due to artificial reefs are on the order 0.05–0.2 kg/m³, indicating that the catches are primarily fishes aggregated by the reefs and greatly exceed the maximum that could be expected from new production.

This discussion is primarily meant to illustrate two approaches that can be used to estimate the relative magnitude of new and aggregated production attributable to artificial habitats, in order to determine how the structures are functioning and their role in fishery management. Each application needs to be evaluated based on the biological and fishery information specific to that application. For example, the growth of oysters and mussels on reefs in eutrophic waters may result in substantial new shellfish production (Fabi *et al.*, 1989). Chapter 3 presents a discussion on the potential of artificial habitats to provide new production as a function of the ecological characteristics of species at the habitat. Also see Chapter 3 for more comparisons of catches and biomasses between natural and artificial reefs.

III. Discussion

Artificial habitats clearly play a role in fishing systems worldwide, and are increasingly employed by fishery and environmental managers in natural resources conservation and planning. Aspects of that role for artisanal fisheries, and fisheries in general, are presented in the following section.

A. Artisanal Fishing

Artificial habitats have proven particularly effective for artisanal applications where fishing effort is relatively low. However, since such structures serve to change the distribution of fishing effort and fishes, they must be viewed within an overall fishery management plan. Their impacts should be considered in a broad socioeconomic context, rather than just in biological terms or changes in CPUE.

Artificial habitats can substantially reduce travel and search time for artisanal fishermen and improve the catchability of their gear. As long as the total fishing effort in the resource is not great enough to result in overfishing, the effects of these structures on the resource are beneficial. Gear competition between fishermen at the artificial reefs and FADs is a potential problem if effort is not regulated, but these structures could also serve to redistribute fishing effort to resolve competition.

Artificial habitats may be useful in closing areas to trawling, to protect juveniles in shallow nursery grounds, and to provide fishing sites for artisanal fishermen using gear that captures mature fish. The deployment of artificial reefs and FADs ideally should be a community project and fishermen should be involved in their planning, construction, and maintenance.

Artificial reefs and FADs built with local materials of opportunity have a certain appeal, but care should be taken to avoid depleting local forests and mangroves, or polluting the environment with inappropriate materials. Longer lasting structures built from properly ballasted scrap tires and concrete may ultimately prove more economical.

B. Fisheries Management and Other Applications

In the presence of heavy fishing effort, artificial reefs and FADs alone may not be economically beneficial. Measures that regulate gear and the fishing effort at artificial reefs and FADs may be required to avoid resource overfishing, user conflicts, and to improve fishery economics. While the literature documents many studies on the ecology at artificial structures, studies on the broader fishery management and socioeconomic impacts of these structures are lacking (see Chapter 7). For progress to be made in understanding the applications of artificial reefs and FADs, scientists and managers must deploy these structures within an overall fishery management plan consistent with the limitations of the particular artificial habitat. Finally, it is useful to view the application of artificial habitats as a decision to allocate space (the site of the habitat) and marine resources to certain user groups. This allocation and the impacts on all user groups should be understood and consistent with the objective of the artificial habitat.

The following list gives some examples of potential applications of artificial habitats that address specific management needs and that take advantage of the way artificial habitats can change the distribution of resources and fishermen, alter gear, and influence size and species harvested.

- When fishery managers wish to reduce fishing effort, artificial habitats may serve as a "bargaining chip" in negotiation. Artificial habitats can create fishing grounds close to port. Such proximity can improve the economics of fishermen by reducing expenses and increasing catchability, perhaps making it easier for fishermen to accept reductions in overall catch.
- When heavy trawling of near-shore nursery areas results in high mortality of juveniles, artificial reefs can be used to close an area to trawlers by creating unsuitable conditions for trawling.

- When one resource is overexploited, artificial habitats can serve to shift fishing effort to another resource. If soft-bottom resources are heavily fished, artificial habitats may be used to shift some fishing effort to coastal pelagic or hard-bottom resources.
- When competition between resource users is a problem, artificial habitats can be used to separate them. In cases of competition between artisanal fishermen and trawlers, artificial reefs can be used to create areas unsuitable for trawling, but suitable for artisanal usage. Sport divers might avoid competition with other types of fishermen for sites by identifying an area unused for fishing, regulating a prohibition of fishing at the site, and then deploying artificial reefs there to create a desirable dive site.

From a biological perspective, artificial habitat may function in one or all of the following ways: (1) to redistribute exploitable biomass, (2) increase exploitable biomass by aggregating previously unexploited biomass, and (3) improve aspects of survival and growth, thereby providing new production. In all three functions, artificial habitats have the potential to alter fishing effort, gear, size of fish at entry to the fishery, species targeted, and catch. The impact of change in fishing mortality on the stock depends on the relative level of exploitation and the rate of movement of the resource to the artificial habitat.

In artificial reef applications, it is possible to estimate the maximum catches from new exploitable biomass due to the artificial reef, and compare this with the actual catches to determine the extent to which the artificial reef is serving as a benthic aggregating device.

More rigorous experimental designs are needed to document the biological impacts of artificial habitat. These designs need to use large numbers of habitat structures to ensure that sufficient statistical power exists to detect impacts in the presence of considerable natural variation typical of many ecosystems. Also, they may require a control site without artificial habitat. Data of a time series should be collected at the treatment and control sites before and after the deployment of the artificial habitat. Fishery-dependent and fishery-independent data should be collected on an appropriate spatial scale and resolution to detect impacts at the artificial and natural habitats. (See Chapter 6.)

Since artificial habitat changes the spatial distribution and density of resources and the fishing effort, standard fishery models, which do not explicitly treat this spatial dimension adequately, may not represent the data fairly. For example, application of the Clark and Mangel (1979) model has proven highly useful for understanding processes at FADs. Further application of this model, along with habitat and diffusion models, should result in more realistic models to evaluate potential impacts of artificial habitat (Mullen, 1989; MacCall, 1990).

References

Ambrose, R. F. 1990. Technical report to the California coastal commission. Section H. Mitigation. Report of the Marine Review Committee, Inc., Santa Barbara, California.

Andree, S., editor. 1988. Florida artificial reef summit, Report 93. Florida Sea Grant College, Gainesville.

Aprieto, V. L. 1988. Aspects of management of artificial habitats for fisheries: The Philippine situation. Pp. 102–110 in Report of the Workshop on Artificial Reefs Development and Management, ASEAN/SF/88/GEN/8. September 13–18, 1988. Penang, Malaysia.

Beddington, J. R., and J. G. Cooke. 1983. The potential yield of fish stocks, FAO Fisheries Technical Paper 242. United Nations Food and Agricultural Organization, Rome.

Beverton, R. J. H., and S. J. Holt. 1966. Manual of methods for fish stock assessment. Part II. Tables of yield functions, FAO Fisheries Technical Paper 38. United Nations Food and Agricultural Organization, Rome.

Bohnsack, J. A. 1989. Are high densities of fishes at artificial reefs the result of habitat limitation or behavioral preference? Bulletin of Marine Science 44:631–645.

Bohnsack, J. A., and D. L. Sutherland. 1985. Artificial reef research with recommendations for future priorities. Bulletin of Marine Science 37:11–39.

Bojos, R. M., and F. J. Vande Vusse. 1988. Artificial reefs in Philippine artisanal fishery rehabilitation. Pp. 162–169 in Report of the Workshop on Artificial Reefs Development and Management, ASEAN/SF/88/GEN/8. Penang, Malaysia.

Bombace, G. 1989. Artificial reefs in the Mediterranean Sea. Bulletin of Marine Science 44:1023–1032.

Brock, R. E., and J. E. Norris. 1989. An analysis of the efficacy of four artificial reef designs in tropical waters. Bulletin of Marine Science 44:934–941.

Buckley, R. M., D. G. Itano, and T. W. Buckley. 1989. Fish aggregation device (FAD) enhancement of offshore fisheries in American Samoa. Bulletin of Marine Science 44:942–949.

Campos, J. A., and C. Gamboa. 1989. An artificial tire-reef in a tropical marine system: A management tool. Bulletin of Marine Science 44:757–766.

Chang, K.-H. 1985. Review of artificial reefs in Taiwan: Emphasizing site selection and effectiveness. Bulletin of Marine Science 37:143–150.

Clark, C. W., and M. Mangel. 1979. Aggregation and fishery dynamics: A theoretical study of schooling and the purse seine tuna fisheries. Fishery Bulletin 77:317–337.

D'Itri, F. M., editor. 1985. Artificial reefs: Marine and freshwater applications. Lewis Publishers, Inc., Chelsea, Michigan.

Fabi, G., L. Fiorentini, and S. Giannini. 1989. Experimental shellfish culture on an artificial reef in the Adriatic Sea. Bull. Mar. Sci. 44, 923–933.

Feigenbaum, D., A. Friedlander, and M. Bushing. 1989. Determination of the feasibility of fish attracting devices for enhancing fisheries in Puerto Rico. Bulletin of Marine Science 44:950–959.

Floyd, J. M., and D. Pauly. 1984. Smaller size tuna around the Philippines—can fish aggregating devices be blamed? FAO INFOFISH Marketing Digest 5:25–27.

Grove, R. S., C. J. Sonu, and M. Nakamura. 1989. Recent Japanese trends in fishing reef design and planning. Bulletin of Marine Science 44:984–996.

Hartwick, E. B., P. A. Breen, and L. Tulloch. 1978. A removal experiment with *Octopus dofleini* (Wulker). Journal of the Fisheries Research Board of Canada 35:1492–1495.

Haughton, M. O., and K. A. Aiken. 1989. Biological notes on artificial reefs in Jamaican waters. Bulletin of Marine Science 44:1033–1037.

Hueckel, G. J., R. M. Buckley, and B. L. Benson. 1989. Mitigating rocky habitat loss using artificial reefs. Bulletin of Marine Science 44:913–922.

Hung, E. W. F. 1988. Artificial reefs development and management in Malaysia. Pp. 27–51 *in* Report of the Workshop on Artificial Reefs Development and Management, ASEAN/SF/88/GEN/8. Penang, Malaysia.

Kakimoto, H. 1984. Ecological study of the Bastard halibut or hirame (*Paralichthys olivaceus*) in the sea referring to the moving of adult fish. Pp. 183–189 *in* Fisheries in Japan: Flatfish. Japan Marine Products Photo Materials Association, Tokyo.

MacCall, A. D. 1990. Geographical population dynamics, harvesting and environmental management. U.S. National Marine Fisheries Service, Southwest Fisheries Center, Tiburon, California.

Marten, G. G., and J. J. Polovina. 1982. A comparative study of fish yields from various tropical ecosystems. Pp. 255–289 *in* D. Pauly and G. I. Murphy, editors. Theory and management of tropical fisheries. Proceedings of the ICLARM Conference 9, January 12–21, 1981. Cronulla, Australia.

McGurrin, J. M., R. B. Stone, and R. J. Sousa. 1989. Profiling United States artificial reef development. Bulletin of Marine Science 44:1004–1013.

Miclat, R. I. 1988. Artificial reef development—the Philippine experience. Pp. 63–86 *in* Report of the Workshop on Artificial Reefs Development and Management, ASEAN/SF/88/GEN/8. Penang, Malaysia.

Milon, J. W. 1989. Artificial marine habitat characteristics and participation behavior by sport anglers and divers. Bulletin of Marine Science 44:853–862.

Mullen, A. J. 1989. Aggregation of fish through variable diffusivity. Fishery Bulletin 87:353–362.

Myatt, E. N., and D. O. Myatt, III. 1990. A study to determine the feasibility of building artificial reefs in Maryland's Chesapeake Bay. Maryland Department of Natural Resources, Tidewater Administration, Fisheries Division, Annapolis.

Polovina, J. J., and I. Sakai. 1989. Impacts of artificial reefs on fishery production in Shimamaki, Japan. Bulletin of Marine Science 44:997–1003.

Prince, E. D., O. E. Maughan, and P. Brouha. 1985. Summary and update of the Smith Mountain Lake artificial reef project. Pp. 401–430 *in* F. M. D'Itri, editor. Artificial reefs: Marine and freshwater applications. Lewis Publishers, Inc., Chelsea, Michigan.

Reggio, V. C., Jr., compiler. 1989. Petroleum structures as artificial reefs: A compendium. Proceedings of the Fourth International Conference on Artificial Habitats for Fisheries, Rigs-to-Reefs Special Session, November 4, 1987, Miami, Florida, OCS Study/MMS 89-0021. U.S. Minerals Management Service, New Orleans, Louisiana.

Samples, K. C. 1989. Assessing recreational and commercial conflicts over artificial fishery habitat use: Theory and practice. Bulletin of Marine Science 44:844–852.

Samples, K. C., and J. T. Sproul. 1985. Fish aggregating devices and open-access commercial fisheries: A theoretical inquiry. Bulletin of Marine Science 37:305–317.

Sato, O. 1985. Scientific rationales for fishing reef design. Bulletin of Marine Science 37:329–335.

Scarborough-Bull, A. 1989. Some comparisons between communities beneath petroleum platforms off California and in the Gulf of Mexico. Pp. 47–50 *in* V. C. Reggio, Jr., compiler. Petroleum structures as artificial reefs: A compendium. Proceedings of the Fourth International Conference on Artificial Habitats for Fisheries, Rigs-to-Reefs Special Session,

November 4, 1987, Miami, Florida, OCS Study/MMS 89-0021. U.S. Minerals Management Service, New Orleans, Louisiana.

Sinanuwong, K. 1988. Artificial reefs construction in Nakornsrithammarat Province. Pp. 130–134 *in* Report of the Workshop on Artificial Reefs Development and Management, ASEAN/SF/88/GEN/8. Penang, Malaysia.

Stanley, D. R, and C. A. Wilson. 1989. Utilization of offshore platforms by recreational fishermen and scuba divers off the Louisiana coast. Pp. 11–24 *in* V. C. Reggio, Jr., compiler. Petroleum structures as artificial reefs: A compendium. Proceedings of the Fourth International Conference on Artificial Habitats for Fisheries, Rigs-to-Reefs Special Session, November 4, 1987, Miami, Florida, OCS Study/MMS 89-0021. U.S. Minerals Management Service, New Orleans, Louisiana.

Stone, R. B., H. L. Pratt, R. O. Parker, Jr., and G. E. Davis. 1979. A comparison of fish populations on an artificial and natural reef in the Florida Keys. Marine Fisheries Review 41(9):1–11.

Sungthong, S. 1988. Artificial reefs development as a tool for fisheries management in Thailand. Pp. 87–91 *in* Report of the Workshop on Artifijcial Reefs Development and Management, ASEAN/SF/88/GEN/8. Penang, Malaysia.

U.S. National Research Council. 1988. Fisheries technologies for developing countries. Report of an ad hoc panel of the Board on Science and Technology for International Development. National Academy Press, Washington, D.C.

Yamane, T. 1989. Status and future plans of artificial reef projects in Japan. Bulletin of Marine Science 44:1038–1040.

Young, C. 1988. Major artificial reef in WA follows SA lead. Australian Fisheries 47(12):26–28.

Environmental Assessment and Monitoring of Artificial Habitats

S. A. BORTONE

Department of Biology
University of West Florida
Pensacola, Florida

J. J. KIMMEL

Florida Department of Natural Resources
St. Petersburg, Florida

I. Introduction

Monitoring and assessment of artificial habitats to determine their characteristics and effectiveness has been of interest from the time that humans first introduced objects into the aquatic environment. While the first assessments of artificial habitats may have been only qualitative impressions of a curious observer, they established a basis and interest for periodic evaluation.

In recent times, the planning and construction of artificial habitats has been directed toward more specific objectives (Bohnsack and Sutherland, 1985; Nakamura, 1985; Sato, 1985; Hueckel *et al.*, 1989; Relini and Relini, 1989). Accordingly, a need has arisen in the biological sciences for more specialized methods to quantitatively assess and monitor habitats to determine if objectives are being met. There also has developed a general acceptance that a quantitative, scientific approach to environmental evaluation will yield the most useful understanding of the influences that affect the performance of man-made habitats (e.g., Bohnsack, 1989; Seaman *et al.*, 1989a). The development of a considerable body of ecological theory has led to a more specific investigation of assumptions that underlie various hypotheses of interest. Within the aquatic sciences, innovative assessment

Artificial Habitats for Marine and Freshwater Fisheries
Copyright © 1991 by Academic Press, Inc.

methods have been and continue to be tested and improved (Andrew and Mapstone, 1987). Concurrently, scientists are developing more useful methods of data analysis.

This chapter presents a review of the assessment and monitoring methodology for artificial habitats. The literature referenced herein, although extensive, is not complete. One should be aware of the vast amount of previously published materials that are available. The answers to many questions may be found there. The literature should always be consulted before any project is initiated. A good place to begin is the annotated bibliography on artificial reefs prepared by Stanton *et al.* (1985). To fully understand the application of various techniques it is necessary to appreciate biotic and abiotic factors and their interaction. Thus, we briefly review characteristics of freshwater and marine environments and problems unique to sampling them. It is essential to know the limitations or special considerations that accompany each technique. Part of the intent of the discussion is to permit more rational decisions about what to assess or monitor and which methods are "best" given the study situation and objectives. Where studies or information on artificial habitats are lacking we draw from the literature on natural reef and related ecosystems.

II. Purpose of Ecological Assessment and Monitoring

The design, implementation, and analysis phases of assessment studies on artificial habitats must maintain a focus on the question(s) that motivated the research. A common concern, for example, is whether the purpose for which the reef was constructed has been achieved. Clear objectives contribute to a rational and scientific approach to habitat evaluation. Assessments that are not goal-oriented can be costly, in view of expenses for personnel and ship operating costs. Further, poorly designed studies can provide misleading and irrelevant information that may lead to inaccurate conclusions.

It is preferable to formulate the objective of any study with reference to a testable hypothesis (Green, 1979; Stewart-Oaten *et al.*, 1986). This means making a generalized statement (based on previous observations) about the topic and then conducting experiments or additional observations that are designed to reject the statement. If a given hypothesis is found totally or partially false, the next step is to identify those parts that are false and formulate a new hypothesis. It may be necessary to modify the questions and redefine the hypotheses until they can no longer be rejected. For example, Spanier *et al.* (1985) posed the hypothesis that recruitment of fishes to an artificial reef might be enhanced by "baiting" it and then compared rates of recruitment to baited and nonbaited sites. Their conclusion was that their hypothesis was not rejected. But had the recruitment rate been equal to or less than that of a nonbaited reef, they would have had to reject their hypothesis and offer a new, modified hypothesis.

Forming a series of hierarchial questions or hypotheses to establish the purpose of an assessment program also determines which methodology is most appropriate for data collection and analysis. In the following sections, we address four principal purposes for assessment and monitoring of artificial habitats, which in turn dictate the selection of methodology to obtain pertinent information.

A. Status of the Artificial Reef

Knowledge of the status of an artificial reef is essential before appropriate questions can be formulated. To determine a reef's "status" some *a priori* knowledge of the reef is needed (e.g., background literature or data from pilot studies). The status should reflect an accurate "picture" of the reef. It may consist only of data from one small place during a brief time. Because of the suite of possible hypotheses to be tested and the natural variability of data, it is often better to have several "pictures" over an extended period. Without knowledge of reef status it would be difficult to formulate hypotheses about habitat size, complexity, carrying capacity, or effects on biomass (e.g., Prince *et al.*, 1985).

The status of an artificial reef is relative to some standard, such as, to itself over time (Hastings, 1979); another artificial reef (Bortone and Van Orman, 1985a); a nonreef area (Alevizon and Gorham, 1989); or a natural reef (Fast and Pagan, 1974). The status of an artificial reef could include information ranging from many variables, measured continuously to only one observation about one variable. Clearly the term "status" is quite broad and the description possible from any reef can vary considerably in complexity and detail.

1. Duration of Study

Various time frames can be used to establish reef status, depending on the study objectives. It is not possible to monitor all habitats all the time but the intervals between sampling surveys can affect whether or not an impact can be detected by the sampling strategy. For example, a short-term survey on a newly established reef (e.g., once a month for a year) may not be long enough to record conditions that may take years to become established. Likewise, a survey done once a year but always on the same tidal cycle may not be able to detect the influence of tides on the faunal composition of a reef inhabited chiefly by tidally influenced species. The temporal basis for studies can be hourly, to account for reproductive or other activity (Colton and Alevizon, 1981; Harmelin-Vivien *et al.*, 1985); night to day, for monitoring fish population changes (Rutecki *et al.*, 1985; Zahary and Hartman, 1985; Bortone *et al.*, 1986; Moring *et al.*, 1989); a month or less, for evaluation of tidal and lunar effects on the biota (Prince *et al.*, 1985; Ardizzone *et al.*, 1989; Bailey-Brock, 1989; Buckley and Hueckel, 1989; Moring *et al.*, 1989; Rountree, 1989); from three months to a year, to evaluate seasonal influences (Bailey-Brock, 1989; Bell *et al.*, 1989); or many years, to account for long-term changes (Buckley and Hueckel, 1989).

Monitoring may need to be conducted over several years to adequately assess colonization and succession. Faunal and floral elements may only gradually colonize new habitat structure (Sutherland, 1974; Sutherland and Karlson, 1977). Turner *et al.* (1969) found no mature fouling communities on an artificial reef in southern California, after more than five years of study. For a general reference to colonization and succession see Begon *et al.* (1986). Readers should be aware that succession is not universally accepted by all contemporary ecologists (see Connell, 1978). (See also related discussion in Chapter 3.)

The inevitable but unpredictable nature of many natural events demonstrates the importance of long-term monitoring. However, it is equally important to note that it may be necessary to adapt the time frame of some surveys to accommodate the short-term impacts of environmental perturbations, caused by natural or artificial influences such as storms or pollution events.

2. Comparative Studies

Comparative studies of artificial reefs are often required by government regulation. Superficially, they seem an easy and straightforward way to determine the effectiveness of a reef. However, environmental assessment conducted over a brief time at a limited locality may not give a representative picture of mechanisms controlling resource availability. Too often, decisions are made based on human biases toward constructing reefs with features that may not be important. This limited view has led to the rapid (and not necessarily justified) expansion of reef construction (Meier *et al.*, 1989). We do not suggest avoiding interreef assessment studies, but they must be planned and the data interpreted carefully.

Comparing the fauna and flora of an artificial reef to a natural reef is an inevitable consequence of the human comparative process and is a worthwhile endeavor. The question of whether or not artificial reefs increase productivity or merely act as resource attractants is entirely in order (Bohnsack and Sutherland, 1985; Bohnsack, 1989). Studies need to examine movement of organisms between natural and man-made habitat to measure the impact of artificial reefs on natural reefs (Brock *et al.*, 1985; Fast and Pagan, 1974; Matthews, 1985). Care should be taken to assure that the assessments are made in an unbiased, scientific, and responsible manner, and are absolutely nondisruptive to the habitat.

B. Relation to Management Objectives

Artificial reefs have had, and will continue to have, an application to fisheries management (reviewed in Chapter 5). Sampling strategy must address the reef's principal management objective (Ahr, 1974). For example, reef placement and design may target a particular species or group of species (Sato, 1985), such as the attraction and habitat improvement of midwater fishes (Stephan and Lindquist, 1989; Buckley *et al.*, 1989), or providing benthic habitat for crustaceans (Davis, 1985). Other questions include, Was the rationale for providing the artificial reef justified? Can the resource base support a potential increase in exploitation rate provided for by the placement of an artificial reef? Was the management strategy met? If the reef was built to increase the maximum sustainable yield (MSY) or optimum yield (OY) for a fishery (Roedel, 1975), then data collection should be designed and scheduled to measure the appropriate factors to monitor MSY over time (Buchanon, 1974; Feigenbaum *et al.*, 1985; Gannon *et al.*, 1985; Solonsky, 1985; Polovina and Sakai, 1989). In another situation, artificial reefs can serve as a refuge for a species and are "off limits" to certain fishing methods, or times of year. Or, they can be used as a refuge to aid in the recovery or maintenance of an existing natural reef community (Brock *et al.*, 1985; Seaman *et al.*, 1989a; Feigenbaum *et al.*, 1989a), or as a means to avoid user conflicts among fishermen. A reef also could provide a safe haven for juveniles, preferred food for adults, a reproductive sanctuary, or enhance survival of a precarious life stage needed to assure a critical level of productivity (Anderson *et al.*, 1989; Campos and Gamboa, 1989).

Nonbiological factors also motivate deployment of artificial habitats. For example, increased local tourism oriented toward SCUBA diving or reduction in user conflicts (Samples, 1989) may be the reason for establishing a reef (Fig. 6.1). When an artificial reef is emplaced to broaden user participation, a much different type of sampling design may have to be considered than that used when the objective of a reef is to increase biomass. Economic and social assessment of habitats is addressed in Chapter 7.

C. Determining the Influencing Factors

To understand and ultimately manage an environmental system it is important to know the specific controlling or influencing factors (Gauch, 1982;

Figure 6.1 Scientific assessment of artificial habitats is often motivated by various fishery interests and concerned with achievement of management objectives. In the northern Gulf of Mexico the gray triggerfish (*Balistes capriscus*) is designated as an "underexploited" species that
can provide a new resource to meet increased recreational fishing pressure in the coastal United
States.

Bortone and Van Orman, 1985a; Patton *et al.*, 1985). It may be possible to
manipulate key environmental variables to improve a habitat's performance
in meeting management objectives. Through experimental manipulation we
may gain insight into the role of various factors in the basic ecology of the
entire aquatic ecosystem (Bohnsack and Sutherland, 1985). To simplify the
research approach, we will review briefly two categories of influencing fac-

tors and their potential significance to artificial habitats. (See Chapter 3.) While not always mutually exclusive, their effects are different enough to warrant separate consideration.

1. Density-Dependent Factors

Density-dependent factors control the growth of a population in such a way that as the population size changes, so does the degree of control of the factor. The relationship can be positive, negative, or in some combination, depending on the interaction of other factors. Density-dependent factors have a quantitative influence on the population with regard to the quantity of something needed by the individuals in the population, such as space or food. For example, a species may respond to the amount of food available as a density-dependent factor (Pitcher and Hart, 1982): the more food present, the greater the potential food consumption, and, therefore, the greater the population. The response by the community may not always be so dramatic or immediate. Concomitantly, there are natural limits to density-dependent factors as population growth rate operates.

Biotic features are often thought of as being density-dependent (e.g., predator–prey relationships). Meanwhile, some nonliving (i.e., abiotic) factors can have a direct (positive, negative, or combinational) impact on an organism's life history as well. If artificial reefs contribute to needed factors in the aquatic environment (e.g., reef size [Grove and Sonu, 1985], number of hiding places or rugosity [Chandler et al., 1985]), then by making those factors more available the reef may be physically enhanced from a human perspective. Ultimately, this is the most significant concept justifying research associated with artificial reefs. The presumption has been that if a species is limited by its habitat, then providing more of the habitat should result in an increase in the fishery resource.

2. Density-Independent Factors

Density-independent factors are conceptually more difficult to understand than density-dependent factors. Density-independent factors are those whose severity and influence is not dependent on the density of the population (e.g., Pennak, 1964). For example, artificial reef assemblages have been impacted by storms (Bortone, 1976; Matthews, 1985) and red tides (Smith, 1975).

Generally, most density-independent factors are abiotic environmental factors. Tides, storms, currents, waves, and substrate are among the density-independent factors that are most often examined and recorded in artificial reef studies (e.g., Lukens et al., 1989). Sometimes it is nearly impossible to separate their effects from those usually considered density-dependent. The phenomenon of upwelling may be an example of a density-independent factor that impacts the amount of available oxygen, but oxygen is a density-dependent factor in most aquatic communities.

Features of reef design, such as surface materials, position, geographic location, configuration, area, and volume, are often significant aspects of the artificial reef environment (Brock and Norris, 1989; Sheehy, 1985). These can be density-dependent if the organisms require them as part of their life history and if abundance is limited or directly controlled by the relative amount of their presence (e.g., number or size of crevices). Similarly, reef design features relative to position (e.g., substrate, distance from shore, depth, amount of light) can act as density-independent factors.

D. Testing Scientific Hypotheses

Artificial habitats often are used as models to experimentally test various hypotheses about the environment (Bohnsack, 1989). Because many of their physical aspects can be manipulated or controlled, they present an opportunity to better understand the aquatic environment (e.g., Hixon and Beets, 1989), especially in conjunction with studies on natural habitats such as coral reefs. Study topics include the species–area hypothesis (Molles, 1978); colonization theory (Fast and Pagan, 1974; Sale and Dybdahl, 1975; Smith and Tyler, 1975; Talbot et al., 1978; Lukens, 1981); species diversity (Slobodkin and Fishelson, 1974; Helfman, 1978); and whether a fauna associated with a reef is present by chance (stochastic processes) or is predictable (deterministic processes) owing to the intimate interaction of environmental factors (e.g., Dale, 1978; Helfman, 1978; Sale, 1978, 1980; Talbot et al., 1978; Sale and Dybdahl, 1975; Smith and Tyler, 1975).

Some studies have tried to manipulate the artificial reef environment to determine which factors are the most limiting to certain species (Hixon and Beets, 1989). By adding or subtracting habitat, competitors, food, parasites, or other factors suspected of controlling life history parameters, it is likely that answers to questions involving these factors will be found while causing little or no damage to natural reefs. Moreover, questions about the utility of artificial reefs for increasing biological productivity can be tested by careful experimental manipulation of the various environmental components of an artificial reef. (See Fig. 6.2.)

III. Problems in Assessing Environments

Monitoring and assessment of artificial habitats must account for certain problems and be cognizant of environmental variation, sampling error, and

Figure 6.2 Much of the experimentation with artificial habitats has involved relatively small structures. Concrete blocks in various configurations commonly are used; see also Figure 1.10. (Photograph courtesy of J. Bohnsack, U.S. National Marine Fisheries Service.)

biotic and abiotic interactions. It is essential to understand what data are to be collected and how they will be statistically analyzed, before beginning actual study. The reader should refer to general texts on environmental sampling and data analysis such as Gauch (1982), Green (1979), and Ludwig and Reynolds (1988).

Some variation is inherent to any factor in the environment, whether biotic or abiotic. Interaction with other variables or features in the environment can reduce or exaggerate the inherent variation that exists (Green, 1979; Poole, 1974; Zar, 1984). Variation also can result from the sampling technique used to conduct an assessment (i.e., the error due to our inability to measure anything with absolute accuracy). This so-called "sampling error" can be reduced with further refinement of technique and careful study.

Environmental research concerning artificial habitats must consider the potential effect of intra- and interspecific competition for available resources. If competition for resources increases to a level that limits individual or species survival, it may influence species abundance or the intensity with which that organism interacts in the life zone. Species interactions usually have a positive or negative impact on the well-being of the species involved (Paine and Levin, 1981). However, determining causative factors and measuring the intensity of the impact is not always straightforward.

A. Assessment Problems in Aquatic Environments

The physical proximity, climate, structural composition, and natural and human-induced perturbations associated with terrestrial environments adjacent to the shore can have direct and significant influence on the aquatic environment. Submerged substrates are often derived from terrigenous sources. Suspended materials including solids, as well as living materials, can greatly impact water clarity and other features. For example, light penetration, water column productivity, and dissolved oxygen content can be interconnected. In addition, gradient (especially in freshwater and near-shore coastal environments) and slope can have the most profound influence on the aquatic community. Variables such as current flow, tidal height and frequency, sediments and deposition rates, and mixing are particularly affected by gradient.

Aquatic ecosystems not only are multidimensional but also present a near weightless environment to organisms. Since water is approximately 800 times more dense than air, a greater variety of materials might be suspended for greater periods of time than is possible in the terrestrial environment. Further, time takes on additional significance, because water column features at one time and place may eventually move to another place. This feature should be considered in any attempt to assess the environment and its resources. A general perspective of aquatic environments is provided by Tait (1981) and Nybakken (1982) for oceans and estuaries, and Cole (1983) for freshwater rivers, lakes, and streams.

1. Assessment Problems Unique to Reef Communities

Assessment techniques must account for the spatial irregularity that reefs present. Reefs are structures composed of a variety of natural and artificial materials. Natural reefs can be composed of living materials such as hard coral, soft coral, sponges, oysters, or vegetation. They also can be composed of dead materials such as submerged logs, sipunculid worm casings and "drowned" coral reefs, or nonliving materials such as rocks, natural deposits, hardened volcanic extrusions, and other geologic formations. All provide substrate heterogeneity and physical relief.

Reef communities are characterized by the very different activities that organisms display. Activity can shift dramatically from day to night (Starck and Davis, 1966; Zahary and Hartman, 1985) or with changes in the aquatic environment such as turbidity or season (Harmelin-Vivien et al., 1985). As the activity of organisms changes, so does the ability of scientists to detect them.

2. Assessment Problems Unique to Artificial Reefs

Human intervention via creation of artificial habitats adds complexity to the aquatic environment. Artificial reefs can be and often are significant modifications to the natural environment. After deployment, an artificial reef becomes interactive with the surrounding habitat. As described earlier both positive and negative interactions can occur. The placement of the artificial habitat, therefore, introduces a source of variation in the natural habitat.

The types of materials from which artificial habitats are constructed may present the assessor with special conditions. They may contribute to water column clarity and purity. Chemicals from the materials may become incorporated into the tissues of reef-colonizing organisms and affect their behavior or other aspects of their life history. The spatial complexity of the reef material may render some assessment methodologies useless, and special techniques may have to be developed for each set of assessment circumstances. Above all, the presence of an artificial reef is likely to alter the species–species, species–environment, and environment–environment interactions found in an area (Fast and Pagan, 1974; Russell et al., 1974; Molles, 1978; Chandler et al., 1985; Alevizon and Gorham, 1989).

3. Assessment Problems Unique to
Artificial Reef Organisms

A fascinating aspect of studying the organisms associated with artificial habitats is the unique role each plays within the biotic community. Unique species mixes and communities (or assemblages, see Chapter 3) may occur. Altered trophic level interactions can result from the altered species mix. Occasionally the presence of a species at a site where it was not found before the deployment of habitat structure indicates that colonization was influenced by the availability of some limiting factor (e.g., space, shelter, or food) not available previously. Although we might argue that a species occurring at an artificial reef is preadapted to the features provided by the reef, the species may have modified its "natural" life history features to be able to exist in its new environment. The assessment problem that this presents is that in the new array of circumstances a species may take on a new or altered role or niche. Sometimes it may be necessary to alter the plan of assessment to accommodate the altered life-style of a species in its new circumstance.

The interactions of organisms and substrate influence other members of the artificial reef community. Bailey-Brock (1989) showed that encrusting species on an artificial reef make the reef surface even more complex with time. It is important to be aware of such factors when designing any study.

IV. Data Needs

The data to be obtained in an assessment program can be identified after the objectives of the artificial habitat have been established, and the limitations that the environment can place on a study are recognized (Caddy and Bozigos, 1985). One approach is to list the features both of an organism that we expect to be affected by habitat deployment, and of the environment expected to have an effect on that habitat. Virtually all variables can be assigned to one of these two groups. These then become the dependent and independent variables. Usually it is the dependent variables we are interested in evaluating as they are impacted by the independent variables. Since artificial reefs represent such an interactive circumstance, several variables may be interdependent or codependent.

Faced with a seemingly unlimited number of variables and types of data, the scientist initially needs to identify those variables best suited to answer the specific questions that have been posed. Some studies will determine the effect of one independent variable on one dependent variable (e.g., the relationship of hole or hiding place size with regard to fish size; Hixon and Beets, 1989). Others incorporate many variables (e.g., ecology of artificial and natural reefs; Bortone and Van Orman, 1985a; Duval and

Duclerc, in Harmelin-Vivien *et al.*, 1985). While a research program may seek to assess the greatest number of variables to the highest degree of precision and accuracy, clearly there are both physical and fiscal limitations to research. The following discussion addresses the types of data that are generally more useful regarding questions concerning artificial reefs. These are more or less "standard" variables that are obtained in many studies. Their utility lies in not only their usefulness in a description, but also their value in comparing data from other studies. The following outline lists variables commonly measured and initial references to methods of assessing and monitoring artificial habitats.

I. Biotic Variables
  A. Fishing
     1. CPUE: Brock, 1985; Crumpton and Wilbur, 1974; Feigenbaum *et al.*, 1985; Matthews, 1985; Myatt *et al.*, 1989; Polovina and Sakai, 1989; Relini and Relini, 1989; Solonsky, 1985
     2. Survey: Crumpton and Wilbur, 1974; Milon, 1989b
     3. Recruitment: Buckley and Hueckel, 1989
  B. Abundance
     1. Number of individuals: Alevizon and Gorham, 1989; Anderson *et al.*, 1989; Bailey-Brock, 1989; DeMartini *et al.*, 1989; Moffit *et al.*, 1989; Thorne *et al.*, 1989
     2. Cover: Bailey-Brock, 1989; Bohnsack, 1979; Fitzhardinge and Bailey-Brock, 1989
     3. Density: Ambrose and Swarbrick, 1989; Anderson *et al.*, 1989; Ardizzone *et al.*, 1989; DeMartini *et al.*, 1989; Harmelin-Vivien *et al.*, 1985; Jones and Chase, 1975; Thresher and Gunn, 1986
     4. Diversity: Baynes and Szmant, 1989; Moffit *et al.*, 1989
     5. Dominance: Jones and Chase, 1975
     6. Community similarity: Jones and Chase, 1975
     7. Occurrence: Anderson *et al.*, 1989; Fitzhardinge and Bailey-Brock, 1989; Laufle and Pauly, 1985
     8. Richness: Moffit *et al.*, 1989
     9. Relative importance: Fitzhardinge and Bailey-Brock, 1989; Jones and Chase, 1975; Laufle and Pauly, 1985
     10. Biomass: Anderson *et al.*, 1989; Brock and Norris, 1989; Jones and Chase, 1975; Moring *et al.*, 1989; Moffit *et al.*, 1989
     11. Colonization: Ardizzone *et al.*, 1989
  C. Life history
     1. Age: Gannon *et al.*, 1985; Moring *et al.*, 1989
     2. Growth: Fabi *et al.*, 1989; Moring *et al.*, 1989; Prince *et al.*, 1985
     3. Size

    a. Length: Ambrose and Swarbrick, 1989; Anderson *et al.*, 1989; Bell *et al.*, 1985; Brock, 1985; Brock and Norris, 1989; De-Martini *et al.*, 1989; Fabi *et al.*, 1989; Feigenbaum *et al.*, 1985; Gannon *et al.*, 1985; Moring *et al.*, 1989

    b. Weight: Ambrose and Swarbrick, 1989; Brock, 1985; Brock and Norris, 1989; Fabi *et al.*, 1989; Feigenbaum *et al.*, 1985; Gannon *et al.*, 1985; Moring *et al.*, 1989

  4. Condition factor: Prince *et al.*, 1985

  5. Feeding: Brock, 1985; Buckley and Hueckel, 1985; Gannon *et al.*, 1985; Jessee *et al.*, 1985; Moring *et al.*, 1989

  6. Habits: Brock, 1954

  7. Preferred habitat: Jessee *et al.*, 1985; Moring *et al.*, 1989

  8. Predation: Jessee *et al.*, 1985

  9. Larval development: Jessee *et al.*, 1985

 10. Reproduction: Gannon *et al.*, 1985; Jessee *et al.*, 1985; Moring *et al.*, 1989

 11. Migration: Buckley and Hueckel, 1985; Davis, 1985; Myatt *et al.*, 1989

D. Abiotic Variables

  1. Substrate: Ahr, 1974; Ambrose and Swarbrick, 1989; Baynes and Szmant, 1989; Chandler *et al.*, 1985; Mathews, 1985

  2. Local conditions

    a. Location: DeMartini *et al.*, 1989; Lukens *et al.*, 1989; Stanley and Wilson, 1989

    b. Temperature: Sanders *et al.*, 1985

    c. Visibility: Ahr, 1974; Carter *et al.*, 1985; Kevern *et al.*, 1985; Sanders *et al.*, 1985; Stephan and Lindquist, 1989

    d. Pollution: Ahr, 1974; Mathews, 1985

    e. Waves and sea state: Mathews, 1985

    f. Weather: Lukens *et al.*, 1989

  3. Reef attributes: Alevizon *et al.*, 1985; Ambrose and Swarbrick, 1989; Grove and Sonu, 1985; Helvey and Smith, 1985; Hixon and Brostoff, 1985; Lukens *et al.*, 1989; Molles, 1978; Sanders *et al.*, 1985

  4. Earth-Tuned features

    a. Lunar cycles: Sanders *et al.*, 1985

    b. Currents: Baynes and Szmant, 1989; Lindquist and Pietrafesa, 1989

  5. Season: Sanders *et al.*, 1985

  6. Time of day: Sanders *et al.*, 1985

A. Biotic Variables

Biotic variables are usually dependent variables. Most often there is some aspect of organism life history, population, or community ecology that is affected by some other biotic or abiotic variable that we can define as important. The complexity of the interactions increases with each higher level of organization.

The species-specific life history parameters can be many and may include life stages (larvae, juvenile, or adult; Grove and Sonu, 1985) or sex (Gannon *et al.*, 1985). Growth data include rate of growth, maximum size, and size at life stage (Gannon *et al.*, 1985; Prince *et al.*, 1985; Ambrose and Swarbrick, 1989; Fabi *et al.*, 1989). Reproductive parameters include age (or size) at sexual maturity, fecundity, reproductive season, reproductive mode (many artificial reef-associated marine species are hermaphroditic) and behavior, and spawning conditions (Jessee *et al.*, 1985). Food and feeding habits as well as inter- and intraspecific behavioral interactions are also important (Jessee *et al.*, 1985; Sutherland, 1974; Fitzhardinge and Bailey-Brock, 1989). Immigration of organisms to the habitat, and the activities of resident or migratory species also are important behaviorial information (Hixon and Beets, 1989). Studies of the habitat requirements of species can be oriented toward the details of microhabitat (fine-scale) or macrohabitat (gross-scale) (Jessee *et al.*, 1985).

Population parameters include abundance (Anderson *et al.*, 1989), sex ratio, and life stage (e.g., Grove and Sonu, 1985). Sometimes more important to our understanding of a population than sheer numbers is the size or biomass of the specific life stages of the population. Biomass present at one time is most often referred to as "standing biomass" or "standing crop" (Bardach, 1959; Brock and Norris, 1989) and is used as a measure of the population (Anderson *et al.*, 1989). A more exacting study would determine biomass in terms of productivity (rate of biomass change per unit time per unit of area or volume). However, such rates are difficult, if not impossible, to measure because of the highly mobile, and thus elusive, nature of the highest proportion of the organisms associated with artificial reefs. However, productivity has been generally inferred from standing biomass (Bohnsack and Sutherland, 1985). Additional population parameters might include migration (immigration and emigration) as well as a host of fishery parameters such as recruitment (i.e., age or size of first capture) and mortality (fishing, natural, and total).

Fisheries associated with artificial reefs do not usually target a single species but often target an assemblage of organisms. This is because of the nonselective fishing methods usually employed and the occurrence of many species susceptible to the gear. Numerous studies have developed stock assessment estimates from catch data (e.g., Grove and Sonu, 1985; Feigen-baum *et al.*, 1989a,b). Ecologically, it is difficult to study individual species without reference to the others with which it associates. Community or species assemblage parameters, therefore, include species composition (species presence or absence), abundance, and diversity (as measured by H', species richness and species evenness; Pielou, 1966). Interspecific associations include commensal, symbiotic, parasitic, and predator–prey interactions.

All biotic components may have potentially significant impacts on the desired target species of the artificial habitat, although thus far research has neglected some of these components (Luckhurst and Luckhurst, 1978). For example, only limited assessment of the attached community, i.e., sessile animals and attached algae, has occurred (e.g., Fitzhardinge and Bailey-Brock, 1989). The water mass flowing past the reef and the near-reef community also contains important communities (Glynn, 1973) and should be an integral part of artificial reef research. (See Fig. 6.3.)

The impact of human activity such as fishing pressure or boat traffic (Harmelin-Vivien *et al.*, 1985; Mathews, 1985) should be examined if it is likely to be an important influence on the dependent variables. For addi-

Figure 6.3 The potential for development of a large population of sessile organisms attached to an artificial habitat is demonstrated by these mussels "fouling" the legs of a petroleum platform. (Photograph courtesy of American Petroleum Institute and Sport Fishing Institute.)

tional discussion of economic and social variables see Milon (1989a,b) and also Chapter 7 of this volume.

B. Abiotic Variables

Many nonliving parameters are studied as part of artificial habitat assessments and are usually considered independent variables. They are essential to examining the impact of the habitat and its surrounding abiotic environment on the biological attributes of the artificial habitat. However, if one is concerned about the biological portion of the artificial reef environment and how it affects the reef itself (e.g., the surface materials after having been encrusted), then the measurements of the abiotic community could be treated as dependent variables. These most often include the more familiar and easily measured water condition parameters, but other nonliving habitat parameters should be considered as well.

1. Habitat Materials, Construction, and Design

The materials from which the habitat is constructed are often important in affecting the local environment of its community. The chemical composition of materials can influence a reef (Woodhead *et al.*, 1985; Fitzhardinge and Bailey-Brock, 1989; Fabi *et al.*, 1989). Many areas have laws regarding the proper treatment of reef materials. For example, automobiles must have oils and other petroleum products removed. Tires are known to leach toxic substances for considerable periods after deployment (Anonymous, 1974). Certain materials such as fly ash blocks and plastics may prove hazardous (although see Woodhead *et al.*, 1985). While leached materials may become diluted in a vast water body, there is the potential of a local negative impact via accumulation of these products in the food chain. Direct interference with the ability of some attached or settling organisms to take-up residence can occur as well (Cairns *et al.*, 1976).

Some material may enhance the attachment of settling organisms and thus increase potential food items available for predators. Some literature describing colonization rates includes discussion of the presumed beneficial feature they have at helping a reef to "mature" and become a "real" ecological community (e.g., Bailey-Brock, 1989; Fitzhardinge and Bailey-Brock, 1989). Although there is little evidence that preferred target fish species consume these attached items (but see Prince *et al.*, 1985), nevertheless, their presence does seem to enhance the aesthetics of the structure and create a "climate" that enhances the fishery. These "nonconsumable" habitat components may, however, provide the necessary structure or features for the consumable prey items of preferred fishes and macroinvertebrates. Surface texture can have a profound influence on the biofouling community and may be just as important as composition material and should, therefore, be recorded in any reef study.

Grove and Sonu (1983) defined and diagrammed the various artificial reefs components as follows: reef unit, a single module; reef set, a cluster of reef units; reef group, an assemblage of reef sets; and reef complex, an arrangement of the reef groups. (See also Chapter 4, including Fig. 4.5.) Each level of complexity can be important in how it influences the biota.

The reef unit is the basic deployed structure of the reef. It can consist of surplus materials such as tires (or clumps of tires if they are deployed as a unit), or a prefabricated module. Several features of the reef unit are important to consider when gathering data. The surface materials should be identified for composition and texture (Hixon and Beets, 1989). The number and size of the holes, openings, and crevices referred to as habitat complexity (Alevizon et al., 1985; Gorham and Alevizon, 1989) or rugosity (Luckhurst and Luckhurst, 1977; Molles, 1978), may be especially significant in creating hiding places for some species or in facilitating the recruitment of certain species or sizes of individuals to the reef (Hixon and Beets, 1989). The population abundance may be directly dependent on the structural complexity (Harmelin-Vivien et al., 1985). Other obvious parameters of the reef unit are unit height, width, area, and volume (Alevizon et al., 1985).

The data-gathering process described for the reef unit is additive and can be expanded to each component of the reef complex. We must, therefore, have data, not only on the reef units, but also on the overall reef configuration. This includes total reef height (the actual height of the reef above the substrate and closest surface depth); total reef volume; interunit distance or scatter (Lukens et al., 1989); proximity to previously existing structures (Ahr, 1974); general dimensions of the total deployment, including shape, orientation to currents, and slope or gradient of the substrate (Baynes and Szmant, 1989). Reef configuration and orientation information, when coupled with data concerning tides, current direction, sediment drift, and topographic orientation may prove useful in predicting optimal deployment of future structures (Bortone and Van Orman, 1985a).

2. Local Conditions

Information about local habitat must include detailed data on water quality and conditions, sediments, and weather. The "standard" water quality measurements include temperature and salinity. With these it is possible to characterize water masses by calculating density. Water clarity, visibility, or turbidity all relate to the amount of light that is able to penetrate to depth. This parameter is important biologically, but it also influences the ability of divers to make observations. The number of organisms potentially observed when using a visual survey is directly related to water visibility (Sanders et *al.*, 1985). Although visibility is most often measured vertically in the water column from the surface, when related to visual surveys it is better measured horizontally at the depth of the sampling survey (Stephan and Lindquist, 1989).

Other water quality parameters important to water column productivity are nutrients (nitrates and phosphates), oxygen (dissolved oxygen), and particulate carbon. Additional parameters could include specific pesticides, heavy metals, or other compounds, according to the objectives of the assessment.

Sediment parameters are useful when assessing or monitoring artificial reefs (Ahr, 1974). These include the size of sediment particles (measured as grain size), type of sediment (based on percent composition of shell, quartz, or other specific materials), depth, and area. Sediment size and type is significant because it can directly impact the decision where to place an artificial structure (Mathews, 1985). Substrate properties will influence the type and abundance of benthic organisms available as food sources. The amount and rates of siltation can impact filter feeding animals and may be especially important when monitoring nearshore artificial reefs (Mathews, 1985). Artificial reef materials may influence local currents and other aspects of flow (i.e., eddies, turbulence) which, in turn, affect substrate characteristics and thus the longevity of the reef itself.

The position of the structure relative to sources of terrestrial water discharge, such as estuaries, bays, and larger effluent discharges should be recorded (Bortone and Van Orman, 1985a). Freshwater river discharge and runoff can influence artificial habitats closest to terrestrial areas (Hastings, 1979). Many studies have shown that there is a species-specific orientation that can be observed on artificial reefs (e.g., Klima and Wickham, 1971; Lindquist and Pietrafesa, 1989). Other studies indicate that currents are important for settling organisms, larval transport, and therefore, juvenile recruitment. Similarly, distances to major currents, the continental shelf edge, a canyon, or some other structure might have a significant impact on the reef community.

Proximity to potential sources of colonizing organisms is important to note, such as the closest seagrass beds, marshes or other nursery areas (Bortone *et al.*, 1988). However, researchers should be aware that "closest" in spatial distance may not necessarily mean "closest" in practical or effective distance, as water currents may make transport from some distant area more likely than transport from a nearby area.

3. Earth-Tuned Conditions

The stage and intensity of predictable currents or tides should be noted because of their influence on the colonization of available substrates (Sanders *et al.*, 1985; Baynes and Szmant, 1989). Variable current patterns, temperatures, and salinities may prove significant. Although the results of such an influence may not be immediately detected by most assessment studies, the comprehensive long-term impact (i.e., over a few decades or more) may profoundly affect local conditions.

C. Number of Samples

Due to the complex nature of environmental parameters, their inherent natural variability, and the large number of potential interactions, the data obtained in ecological studies should be interpreted with caution. The important questions, What can this variation tell us? and How do we account for variation? must be considered. The answer to the first is rather simple. We can examine the relationship between parameters and compare variability. This usually is done graphically and statistically, as an initial step in understanding the causes of variation and their influence.

Accounting for the sources of variation is another matter. The first step is to acknowledge that variation can and does occur, due to sampling technique, the inherent variation in all things, or some other factor. To describe and account for variation a large number of samples measuring the variables of interest must be taken. The natural variation associated with reef organisms is compounded by the spatial heterogeneity of reefs (Harmelin-Vivien *et al.*, 1985).

The number of samples required varies for each situation (Andrew and Mapstone, 1987). For samples in which the factors are slightly variable relative to their mean value, as few as three to ten replicates may be required. Conversely, more than 50 replicates may be necessary if variance greatly exceeds the mean value of the parameter. The reader may wish to refer to texts such as Green (1979) or Zar (1984), which outline the statistical methodology needed to determine the number of replicates.

Generally, it is reasonable to expect accountability in variation of ecological parameters to occur with 20 to 40 replicates. Ideally more should be taken, but the expense and time usually cannot be afforded. In fact, few artificial reef studies have been conducted using more than five samples to account for variation; of those studies only a few have actually calculated the number of samples necessary. Sale and Douglas (1981) determined that an optimal sample number of four would provide enough information on the variability on reef fish assemblages. Harmelin-Vivien *et al.* (1985) thought a replicate number of at least 12 samples was necessary. Bortone *et al.* (1989) indicated that values of attributes such as species diversity and cumulative number of species stabilized after 16 to 32 samples, depending on the methodology. Bortone *et al.* (1986, 1989), Jones and Thompson (1978), and

Kimmel (1985b) used a minimum of eight samples to account for 90% of the inherent variability using the species–time random count visual survey method based on sample size estimates established by Gaufin *et al.* (1956).

In our view, most studies merely make a guess as to what seems a reasonable (or affordable!) number of replicates. It is not that most scientists are not aware of the importance of having a complete understanding of the nature of variation in artificial reef environments, it is more a limitation placed on the study design by the available equipment, personnel, and other resources. Consequently, the conclusions attained by an assessment and monitoring study should not always be considered as totally accurate and should be interpreted with caution. The inability to adequately account for variability may render some statistical analyses useless.

D. Frequency of Sampling

Frequency of sampling also should reflect the objectives of the study. To assess daily activity of a species on an artificial reef, sampling may have to be done continuously or at least hourly. Day–night activity (e.g., Hobson, 1965; Samples, 1989) in the environment should be monitored at least four times a day (day time, night time, and during the dusk and dawn crepuscular periods). Lunar influences on the artificial reef community are also well known (Hastings, 1979). To account for this variable, sampling would have to occur at least every seven days. A monthly sampling regime generally misses lunar influence. Seasonal variation on the reef should be monitored at least every three months in temperate areas. In some parts of the world, such as the tropics, seasons are more often related to the amount of rainfall and, therefore, should coincide with those regional conditions.

If the prime consideration in assessing an artificial reef is the colonization or defaunation rates, then the sampling design should be at a frequency to evaluate these rates in a meaningful way (Ardizzone *et al.*, 1989; Lukens, 1981). Studies of succession, as indicated above, should be of shorter intervals initially (e.g., weekly, monthly) and later reduced to a seasonal or even yearly monitoring regime, as appropriate. For example, Ardizzone *et al.* (1989) indicated that the succession of a benthic community associated with an artificial reef in the Mediterranean Sea had not become fully stabilized at the climax level after five years. Turner *et al.* (1969) suggested a similar situation in fouling communities off California.

The sampling schedule is especially critical if the objective includes time-related trends. Here sampling should occur at regular intervals and not be missed (Witzig, 1983). Unfortunately, this is almost never possible due to problems with equipment, personnel, weather, or disruption in funding. Additionally, it is important to be flexible in the sampling regime to examine the impact of unusual influences. Bortone (1976) was able to sample an artificial reef a few days before and after a hurricane to describe the storm's impact on it.

V. Assessment Methods

This section reviews the methods used to assess and monitor an artificial reef. There are advantages and disadvantages to each technique, under preferred and nonpreferred situations. Also, methods may have to be modified to accommodate the local situation or the specific study objective. Where most of the methods have been applied to artificial reefs, some (or special features of them) have been employed only on natural reefs. Since faunistically and topographically (see Chapter 3) these habitats present nearly equivalent sampling situations, there should be little difficulty in transferring any method used on natural reefs to artificial reefs. Sampling techniques that are least disruptive to existing conditions reduce the impact of variability induced by sampling and enhance the probability of reliable replicate samples.

Regardless of the environment, each method must be described clearly and referenced so that it can be duplicated. This will aid future research and permit better evaluation of the appropriateness, extent, and quality of the data. Assumptions associated with data collection should be stated (Andrew and Mapstone, 1987). Also, the perceived beneficial and detrimental features of the methods used, and recommendations for their modification should be identified.

A. Assessing Biotic Variables

Special care should be taken not to stress organisms when gathering data. This risks adding a new source of variation to the natural variability already present. The biological adage that an organism's response to stress can be to "adapt, move, or die" should be heeded.

1. Individual Parameters

Sampling design is often aimed at gathering information from individuals first, and later the whole population. Below we discuss several methods which, although aimed at the population and community level of organization, begin by gathering data from the individual organism. These data can be combined easily to form larger units. However, to gather data at a higher level and then reduce them to a smaller unit is nearly impossible. (See Ricker [1968] for a general introduction to methods for assessing life history features of fishes, which also can be applied to macroinvertebrates.)

Data may be gathered through capture of individual organisms. Capture techniques should be nondestructive to the organisms and nondisruptive to the habitat. It is necessary to know if the capture method will bias the data. For example, we have noticed that trap-caught specimens can give an unrealistic picture of feeding habits of some fishes because they tend to feed in the trap, sometimes on organisms that they rarely encounter. If the focus of the study is species specific (e.g., the growth features of a species on two different reefs), it may be necessary to use a collecting technique best suited for that particular species.

It is desirable to return captured organisms to the water quickly, with as little trauma as possible. The degree of handling and amount of time to avoid trauma varies with each species and life stage. Clearly, there is some depth, and rate of retrieval from depth, beyond which survival for some organisms is considerably diminished, owing to rapid dissolution and expansion of gases (Gotshall, 1964). If handling is not excessive, depth is not unreasonable, and retrieval rate is slow, the specimens should survive tagging and release for migration studies (Davis, 1985; Matthews, 1985; Solonsky, 1985; Hixon and Beets, 1989). Survival is unlikely for fish in which the branchial arteries have been ruptured ("gill-hooked"), or that have been dragged by bottom trawls, or captured by other actively fished gears.

Live specimens can provide much information about size, species composition, feeding (regurgitation of stomach contents from live organisms is quite successful), growth, and reproductive condition. If the animals are to be sacrificed, it is essential to optimize the data obtained from each individual. Possible uses include otoliths or bones for growth studies; parasites for levels of intensity and infestation, as well as stock identification; food and feeding habits from stomach contents (also indicates the habitat in which they have been); gonads, for fecundity, maturation, and sex ratios; and various tissues for stock identification.

Information such as reproductive and feeding behavior, local movements, territory size, competitive interaction, and habitat requirements can be gathered through remote or surface-tended collection methods. However, many of these data are best gathered through *in situ* inspection by a diver, through some remote video camera facility, or even a research submersible (Fig. 6.4).

2. Population Parameters

For the purposes of this chapter the terms stock, unit stock, and population are used synonymously. A population (i.e., " . . . a number of genetically similar individuals living in a limited framework of time and space," Emmel, 1976:89) is the sampling unit to which many biological studies are oriented. The term stock is generally considered synonymous with the term

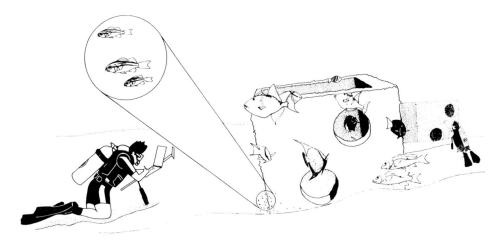

Figure 6.4 The scientist's ability to make observations and record data underwater has greatly enhanced environmental assessment programs concerned with artificial habitats. (Drawing courtesy of J. Bohnsack.)

population (Pitcher and Hart, 1982). During the past two decades there has been an emphasis on "unit stock," an operational concept used by fishery biologists and considered to be a population that can be managed as a single group (i.e., their life history features are similar enough with regard to how they interact with the fishery). Sometimes a unit stock may be a local group of several species that can be managed as a unit because the fishing techniques used to capture them do not permit a separate fishery for each species. For example, in warmer Gulf of Mexico and Caribbean waters, reef fish are generally considered by U.S. fishery managers as a single unit stock even though ten or more species from at least three families (generally snapper [Lutjanidae], grouper [Serranidae], and grunts [Haemulidae]) constitute the fishery. For a general discussion of unit stock see Gulland (1983).

Sampling from a population should be unbiased, or at least the bias should be recognized. In other words, data should be gathered from representatives of the population that were not selected for any particular predetermined attributes (e.g., size). Although it may never be possible to obtain a truly unbiased sample from a population, due to gear selectivity or behavior of the organisms, it is still our goal. Most statistical analyses have a precondition that the data are unbiased and were collected "randomly" from a larger population.

Although specific population analyses presented in fishery textbooks are beyond the scope of this chapter, we summarize methods used to measure or assess relative abundance of the population. The purpose of most population studies is to determine population density, either by species or species group, stated in terms of size or mass per unit area. It is important to calculate the mean value of these parameters as well as their variance (Thresher and Gunn, 1986).

B. Nonvisual Methods for Fishes and Invertebrates

A broad variety of nonvisual techniques are useful when sampling in areas of low visibility (Bardach, 1959; Moring et al., 1989) or when other conditions such as local weather and sea state hinder or prevent visual surveys. This is generally the situation in many freshwater lakes and streams (Crumpton and Wilbur, 1974). Techniques for sampling fishes include hook-and-line, nets (gill, trammel, trawls, etc), electrofishing, traps, creel surveys, and expert angler evaluation (Crumpton and Wilbur, 1974), as well as hydroacoustic techniques (Thorne et al., 1989). Except for trawling, these methods are relatively nondestructive to habitat. Except for creel survey techniques, these have the advantage in artificial reef studies of providing individual specimens for further analysis.

Perhaps the most often used nonvisual sampling method is hook-and-line. It provides specimens for life history analysis, is relatively nondisruptive to the habitat, and often reflects the usage a reef receives (Moring et al., 1989). This is especially important since artificial reefs are frequently built to improve fishing. Obviously, an excellent way to evaluate reef effectiveness is to fish the reef with gear typical to the area.

One effective way of monitoring the fishing status of a reef is to record the catch data in terms of the effort employed, as catch per unit effort (CPUE). Catch can be recorded as number of fish (often as the preferred species sought; Buckley and Hueckel, 1989) or biomass (Brock, 1985). Effort units vary according to the fishery (e.g., numbers of hooks fished per hour, fishermen or boats). CPUE has been recorded, for example, as number of fish per rod hour (Feigenbaum et al., 1985), number of fish and number of strikes (by species) per unit time (Beets, 1989), and biomass per line per unit time (using a fixed number of hooks or lures; Buckley et al., 1989). It is important to consistently define both variables being measured (Matthews, 1985).

Gill and trammel nets are used to assess the standing stock of fishes associated with artificial reefs (Gannon et al., 1985; Relini and Relini, 1989). Usually the same standard gill net used in a commercial fishery is used in scientific sampling, but often an "experimental" net (with varied mesh sizes) is used to capture fish of different sizes (Liston et al., 1985; Moring et al.,

1989). Vertically fished gill nets sample the entire water column (Moring *et al.*, 1989). Effort data from nets also can be recorded to make CPUE estimates (Gannon *et al.*, 1985).

Baited fish traps have been used somewhat effectively (Bardach, 1959; Miller and Hunte, 1987; Bortone *et al.*, 1988). Traps also can provide accurate CPUE data, but only if they are standardized and records are maintained on the amount of time each trap is effectively fished. Moreover, they should be calibrated with an independent survey method such as a visual inspection with SCUBA (Miller and Hunte, 1987). When using unbaited traps, Stott (1970) indicated that estimates on fish abundance may be unreliable unless similar refugia habitats exist throughout the area being sampled.

Data on catch and effort may not always be from a research effort or a fishery-independent study as described previously. Sometimes CPUE data can be from the fishery through a creel census, survey, questionnaire (Buchanon, 1974), ship logs, or fishing tournament catch records (Stanley and Wilson, 1989). Biases can result from poor recall or memory.

Mark-and-recapture techniques are often used to estimate fish and macroinvertebrate populations (Robson and Regier, 1968). Their value is limited in most studies on artificial reefs because of the migratory nature of many species and the difficulty in obtaining large numbers of organisms to mark or tag to obtain a reliable estimate of the total population. Also the low probability of survival of fishes collected at depth, handled at the surface, and returned to depth, creates a strong bias toward overestimatating population size. High variability of data makes this technique unattractive for most artificial reef situations.

In some areas, the historical use of explosives to capture fish in artisanal fisheries has been adopted in scientific studies (Bardach, 1959; Goldman and Talbot, 1976). This practice, too, has bias. For example, sharks have been attracted to the noise and also to dead and dying organisms; they not only remove important specimens from the water column but also may affect the recording performance of the researchers (Russell *et al.*, 1978). Explosives are clearly detrimental to the habitat and some species are more susceptible to them. Explosives, however, overcome the limitations caused by irregular substrates, which plague most net and some hook-and-line sampling methods. Ichthyocides such as rotenone have been used in both artisanal fisheries and scientific sampling (e.g., Smith, 1973). Poisons, however, are not considered an optimal sampling method due to their interference with recolonization as well as the lack of control over their potency and impact on nontarget organisms.

The disadvantage of nonvisual sampling is that some species, or age groups within a species, are biased for or against capture. For example, some species notably seek out traps (Bardach, 1959; Miller and Hunte, 1987), and some can be attracted to areas with various baits or baited hooks (Somerton *et al.*, 1988).

C. Visual Methods for Fishes and Macroinvertebrates

Reliable data on species identification, abundance, size, and distribution may be obtained by an observer, either operating as a diver (either skin or SCUBA) or using a submersible (Moffit *et al.*, 1989; Shipp *et al.*, 1986) or ROV (remotely operated vehicle; Greene and Alevizon, 1989; Van Dolah, 1983). Assumptions that apply include the following: species or species groups can be accurately identified (Brock, 1954; Harmelin-Vivien *et al.*, 1985), numbers of individual organisms (and schools) and their body sizes can be accurately estimated, and the proportions of these organisms counted are representative of the entire habitat (Brock, 1954).

Visual techniques do not disturb the habitat (Bardach, 1959; Harmelin-Vivien *et al.*, 1985) and are minimally disruptive to the organisms. Since they do not remove organisms from the environment they can be repeated to achieve replicate sampling. When SCUBA or skin diving are used, field gear requirements are few. Finally, visual techniques are less selective when compared to most other sampling methods (Brock, 1954).

Visual techniques are flexible and can be adapted to a variety of different situations and habitats, such as benthic and midwater situations (Rountree, 1989). They can provide qualitative data on the condition of an artificial reef (Prince and Brouha, 1974; Bortone, 1976; Myatt *et al.*, 1989), presence of organisms and community structure (Helvey and Smith, 1985), and quantitative data on the density and relative abundance of species or the entire community (e.g., Kimmel, 1985b; Dennis and Bright, 1988).

A special feature of visual assessments involves the ability to accurately record *in situ* observations for various organisms. This ability may be limited by organism mobility; their secretive, sedentary nature; or their coloration pattern (cryptic to conspicuous). Some species vary temporally in color pattern and activity, creating a problem of unequal detection (Brock, 1982; Reese, 1975). These behaviors may violate the assumption that each species and individual has an equal probability of being detected (Harmelin-Vivien *et al.*, 1985). The presence of the observer can also influence the behavior of some species (i.e., attraction or dispersal) thereby affecting detection (Collette and Earle, 1972). Our perception is that diver influences on the fish fauna are minimal because the few species that are affected seem to acclimate quickly to an observer's presence (except where spearfishing occurs). Further study of this aspect may allow the data to be adjusted for bias created by unequal detection between species, sexes, or life history stages.

Disadvantages of visual surveys should be recognized and, if possible, compensations applied to enhance the utility and validity of observations. Inadequate visibility is one of the most common limitations (Fig. 6.5). DeMartini *et al.* (1989) indicated that visibilities below 3 m drastically compromise results. Some methods require greater visibility (e.g., 5.6 m in Bortone *et al.*, 1989). An inverse relationship between object–observer distance and observability was reported by Harmelin-Vivien *et al.* (1985). It has been noted (e.g., Sale and Sharp, 1983; Sanders *et al.*, 1985) that the volume of water that can be effectively surveyed is a function of water visibility. Two common and contrasting protocols exist. In one, observers visually search a predefined area or volume of water (Fig. 6.6). In the second, a variable volume, extending to the limits of underwater visibility is searched. Where visibility fluctuates, error in detection of objects can be magnified and limit data utility. Depth also limits the amount of time spent surveying. It is essential to refer to appropriate dive tables or a diving officer to avoid physio-

Figure 6.5 Much of the methodology employed in visual assessment of artificial habitats has been developed in natural reef systems featuring relatively high water clarity.

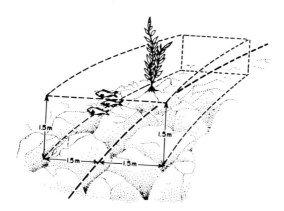

Figure 6.6 A representative diver transect, denoted by heavy line, over benthic habitat surveys a predetermined, three-dimensional horizontal and vertical distance. The boundaries are denoted by dashed lines. (From Jessee *et al.*, 1985.)

logical problems associated with SCUBA diving [National Oceanic and Atmospheric Administration (NOAA), 1975; Lang and Hamilton, 1989]. Strong surges and waves can hinder diver surveys (Spanier *et al.*, 1985). Nocturnal studies obviously require illumination but may bias the fauna recorded due to species being attracted or repelled by lights (Zahary and Hartman, 1985). Use of a red filter between the lens and the light source can reduce the impact of the typical incandescent light spectrum on organism behavior.

Perhaps the most severe limitation of visual faunal estimates is their underestimation of real abundance (Wickham and Russell, 1974; Harmelin-Vivien *et al.*, 1985; Buckley and Hueckel, 1989). This is related to the combination of spatial heterogeneity of the substrate of artificial habitats and the cryptic nature of many organisms (Brock, 1982; DeMartini *et al.*, 1989), which can reduce any individual organism's probability of being observed and result in underestimates of population density (DeMartini *et al.*, 1989).

Conceptually there are differences between the method used to gather data and the method used to record them. Data should be recorded as observations occur, especially in the case of underwater assessments, unless safety is compromised. Rarely is there any advantage to delay recording diver-observation data. However, Rutecki *et al.* (1985) indicated a preference for recording data upon completion of the survey. Obviously, increased delay (even minutes) between making an observation and recording it increases the probability of error.

Plastic slates of various types (Helfman, 1983) are used for most data recording. Typically, data are written directly on the slate, usually with

Figure 6.7 A pencil and a plastic slate, roughened with sandpaper, is a commonly used data-recording method. This diver has prelisted the species likely to be seen on the slate, but many researchers think this biases the sampling survey.

pencil (Fig. 6.7). Plastic (e.g., Mylar) or waterproof paper is sometimes af-fixed to the slate. Slates are inexpensive, and adaptable to each survey. A dis-advantage is that valuable observation time may be lost if the divers must look away from their observations to see what they are recording. Photocopies of original data sheets can serve as a "back-up" and should be kept separate from the originals until the data are stored in a computer data base. There also should be a duplicate of the computer data base.

The best recording methods are those in which data are recorded only once, in a format that is immediately archivable. This reduces the chance of error due to transcription. Some type of data recorder, such as that found on various computer input devices (i.e., diskette or tape media), is probably the best overall method. There have been only a few such studies that have used the technique to date (Hixon, 1980). Most certainly, direct computer input of data will be the mode of *in situ* data recording in the future (see Williams and Briton, 1986).

Alternatives to direct computer input are audio and video recording de-vices that make a permanent copy of the records in a format that is readily transcribed to an archive. These techniques are used both above and below the water (Alevizon and Brooks, 1975; Jones and Chase, 1975; Larson and DeMartini, 1984; Bortone *et al.*, 1986, 1989; Ebeling *et al.*, 1980; Greene and Alevizon, 1989). Audio- and videotapes produce a reasonably permanent record, allow recording of many data items rapidly, can be reexamined to verify an observation, and gather data initially not considered important. Moreover, they represent a relatively inexpensive way to inform nontechnically oriented individuals about a specific study or a data collection method.

The disadvantage to this system is that all data must be transcribed from the film or magnetic tape. Transcription time is usually equal to or greater than the time required to make the recording. Also, these devices are expensive, of varied quality, and require some special training (such as the use of a full-face mask for audio recording) and handling (audio and video recorders are easily damaged by moisture). The limited resolution and field of view of video recorders (not only but especially in murky or poorly lit waters) may preclude accurate species identification (Seaman *et al.*, 1989b) and abundance estimates.

Still photography has been used successfully in data recording (see review by Weinberg, 1981). It has the advantage of being relatively inexpensive and easy to operate (Fig. 6.8). Photography has been used to record relatively sophisticated data on populations and geographic conditions (Lundälv, 1971), but in some instances it produces notably unreliable data, especially under poor visibility. Fishes are extremely difficult to identify or even see from a still photograph, as their movement against a background is often essential for recognition (Bortone *et al.*, 1986). Photographic techniques are discussed in the section on sessile communities below.

Remote recording devices have a definite advantage at depths greater than those safely accessible by SCUBA for some types of data. ROVs and manned submersibles are usually equipped with audio and video recording, either directed to the surface or recorded *in situ*. Automatic depth recorders and positioning systems are invaluable for determining the exact position and extent of some reefs. Recent use of transponders to produce a three-dimensional-like image of submerged objects shows tremendous potential for locating, identifying, and assessing the form of benthic artificial structures (Lukens *et al.*, 1989).

In the following sections we present a more specific discussion of the attributes of visual methods. Readers also are advised to consult Harmelin-Vivien *et al.* (1985) for a comprehensive introduction.

1. Observer Limitations
Most visual assessment methods use a diver as the observer and several limitations and factors can affect performance. Diver ability is an obvious

Figure 6.8 Still photographs are a valuable way to document the reef area for later comparisons, as well as for showing the reef to interested individuals, but they do not permit precise quantitative data acquisition on the nektonic community.

variable that must be considered. Divers unfamiliar with diving and, therefore, preoccupied with diving techniques will be diverted from their study objective. Limited visibility, turbulence, or extreme depth may exacerbate problems among even experienced divers. A potential source of error is the observer's ability (usually related to experience) to accurately count, measure, record, and identify species (Stephan and Lindquist, 1989; Harmelin-Vivien et al., 1985). Narcosis at depths greater than 30 m can occur and lead to inattention, but this can be mediated to some extent by using Nitrox (a mixture of nitogen and oxygen) as a breathing gas (Seaman et al., 1989b). Exposure to cold and diver fatigue can cause disorientation and lack of concentration (Harmelin-Vivien et al., 1985). Participation of an individual assigned the role of "dive safety officer" is essential.

Divers can influence the behavior of organisms. For example, Harmelin-Vivien et al. (1985) noted lower fish abundance when successive visual surveys were conducted over the same area. Kimmel (1985b), however, could detect no difference in abundance of fishes when replicate surveys were conducted along a transect. Some organisms are attracted or repelled by features such as lights and bubbles (Smith and Tyler, 1972). However, the use of a dive light (even diurnally) can facilitate the identification of organisms in dark recesses, at depth, or under other dim light conditions. Often data are taken by both divers operating in pairs for safety purposes. Divers should have the same skill levels with regard to species identification, enumeration estimation, and size judgement. Several studies have evaluated variation among observers. Beets (1989) observed that divers were similar in most aspects of data recording except when surveying uncommon or cryptic species. Bortone et al. (1989) indicated disagreement between two divers with regard to some parameters, such as number of species and species diversity, when using some visual census methods but not with other methods. Epperly (1983) found that even though divers agreed on identification and number of species, individual observer variation was considerably less than between-observer variation for enumeration of individuals.

Another source of variation among observers is attributable to their visual perception ability. Morrison (1980) indicated considerable individual difference for perceiving objects and ability to see moving or stationary objects. Individuals also differ in ability to see objects of different contrast (Ginsburg and Conner, 1983) and to perceive dimensions of objects (Enns and Rensink, 1990).

Clearly when multiple observers are recording visual census data it is necessary to measure their comparability, especially since some observers may produce comparable results using one visual survey method but not others. In any case, diver training seems to improve agreement both within and between observers (Galzin, 1985, in Harmelin-Vivien et al., 1985).

2. Species Identification

Correct species identification may be the most important part of a visual census data base (Harmelin-Vivien et al., 1985). An error here may render even the simplest of species presence or absence analyses completely useless. This procedure may require much training, particularly in areas of high species diversity (e.g., coral reef communities in the Caribbean may potentially have over 400 fish species in residence). Skill is necessary to be able to accurately record the identity of many species of different sizes, swimming behavior, and color patterns. Some species may have separate adult male and female color patterns with different patterns as juveniles, and at night (Starck and Davis, 1966). There may be similarity of color patterns among species, also, due to adaptation or genetic similarity.

To avoid identification mistakes, some studies have eliminated small, cryptic species that have a high probability of being misidentified or missed completely (e.g., Greene and Alevizon, 1989). This is especially practical when the objective is to assess a target species or group of species (e.g., Buckley and Hueckel, 1985; Jessee et al., 1985). Species may be included as

members of a group such as a genus or family, or given a numerical desig-
nation that at least separates one species from another (e.g., Bortone et al.,
1986). A photograph or capture of a voucher specimen can facilitate identification. Spanier et al. (1985) used a trammel net to collect specimens to
verify species and sizes. One of us (J.J.K.) has used quinaldine (an anes-
thetic) and a "micro-barb" spear to collect voucher specimens.

Humans are thought to rely on a "search image" for sight identification
(Enns and Rensink, 1990; Corbetta et al., 1990). During a survey it may be
preferable to make a visual sweep looking for only species with a similar
body profile, size, life stage, behavior, color, position in the water, or relationship to the substrate (Brock, 1954; Molles, 1978; Grove and Sonu, 1985;
Bohnsack and Bannerot, 1986; Moffit et al., 1989). Greene and Alevizon
(1989) referred to these search image groups as discrete census groups. This
strategy may affect the overall sampling if additional time is required for
repetitive visual sweeps for each census group.

3. Enumeration

The enumeration of species (or species groups) also requires some special attention, particularly in view of the extreme range in abundance and
patchy distribution that may occur on both natural and artificial reefs. It is
relatively easy to accurately count a few individuals of a few (target) species
in relatively clear water. Usually, however, conditions are less favorable.

To reduce enumeration errors several things should be considered. (See
Fig. 6.9.) Species enumerations generally underestimate the actual number
present because of the cryptic and secretive nature of many species (Sale and
Douglas, 1981; Sale and Sharp, 1983; Harmelin-Vivien et al., 1985). Wickham and Russell (1974) found that for pelagic species around midwater structures there was a tendency for divers to overestimate small populations and
underestimate larger populations (see also Bevan et al., 1963; DeMartini
and Roberts, 1982). Harmelin-Vivien et al. (1985) found that divers had little
difficulty in correctly enumerating schools of fish of six or seven individuals
but the error of enumeration increased at higher school densities. To deal
with this problem several studies have used abundance groups. Observers
form a search image of a group comprised of a certain number of individuals
and then determine how many groups they see. The group size varies with
each study but generally ranges from 20 to 1000 in size (Harmelin-Vivien
et al., 1985). Instead of counting individuals, Workman et al. (1985) counted
in abundance units (i.e., 1, 1–20 individuals; 2, 21–50; 3, 51–100); Russ
(1984) and Kingsford (1989) used base 3 logarithms; Sanders et al. (1985)
enumerated particularly abundant schools by orders of magnitude (i.e.,

Figure 6.9 In areas of high species diversity and abundance, enumerations can become somewhat complex. For example, an artificial reef in the Bahamas held large populations of sailor's choice, *Haemulon parrai* (foreground), and three surgeonfish species, *Acanthurus* spp. (background).

1×10^2, 1×10^3); Gladfelter *et al.* (1980) used doublings for abundance categories. Other researchers have used words to describe abundance units (i.e., rare, 1; occasional, 2–5; frequent, 6–10; common, 11–25; abundant, > 25; Bortone, 1976). Some studies have used abundance categories but have not clearly defined them (e.g., Hastings *et al.*, 1976; Smith *et al.*, 1975).

Counting by abundance units may considerably facilitate the enumeration process and may lessen the chance of error. Since the species abundance data are often transformed for statistical analyses, recording data in this fashion approaches a log transformation. The chance for error when counting by abundance units comes from incorrectly assigning a group to an abundance class when the group is close to the boundary of a unit (e.g., assigning a group of 25 individuals to an abundance unit of 26–50 individuals; Harmelin-Vivien *et al.*, 1985).

When counting a school of mixed species it is preferable to estimate the number of the entire school and then estimate the proportion or percent composition that each species makes to the school. We have found that mixed-species schools are usually composed of similar appearing species of similar size making species specific enumeration difficult, especially in conditions of poor visibility.

4. Size Estimation

Size estimates are often recorded to evaluate the importance of a species or species group at an artificial habitat. (Additional discussion of the rationale for such data is provided in Chapter 3.) Generally, size is estimated as length and later converted to weight or biomass (Brock and Norris, 1989) using data from studies on length–weight relationships (e.g., Dawson, 1965; Bohnsack and Harper, 1988). Bohnsack and Bannerot (1986) used a rule mounted on a stick and held at a distance to aid in estimating fish body length. Others have used rulers, calibrated slates, surveyor's tape, or some other object of known size as a reference (McKaye, 1977). It is important to have a reference of known length due to the magnification effect of water. Bell *et al.* (1985) and DeMartini *et al.* (1989) found that with practice observers could reliably estimate lengths underwater.

To avoid making small errors in estimating length, some studies have estimated fish size and assigned them to size groups. Harmelin-Vivien *et al.* (1985) classed fish as being < 12 cm long, 12–18 cm, or > 18 cm; Matthews (1985) used a slightly different size grouping: < 6 cm, 6–12 cm; 12–20 cm, > 20 cm. Beets (1989) assigned fish to 2.5-cm units; McCormick and Choat (1987) used 5-cm length classes, and Bell *et al.* (1985) assigned fish to 10-cm length units. Our experience indicates that even inexperienced divers have little difficulty in correctly assigning fish to 5 cm size groups.

Schooling fish do not present a major problem when estimating lengths since fish of equal size tend to school together so that it is not necessary to estimate the size of all individuals in the school but, only one or a few members. Upon occasion different sized individuals may occur within a school. The human eye has the ability to recognize odd individuals (with regard to size or species) in a school, making them relatively easy to detect (Theodorakis, 1989).

5. Visual Census Methods

A plethora of census methods have been employed, with variations peculiar to each investigator or modified to accommodate special circumstances including environmental conditions, study objectives, ability of the investigators, and availability of resources. While flexibility is the greatest advantage of visual census methods, directly comparing data obtained by different methods is difficult if not impossible. Below, the methods are presented by major type with regard to whether or not the observer moves in a predetermined direction (transect), remains stationary (point count or quadrat), or swims haphazardly (species–time random count). Some methods do not fit easily into any category.

   a. Transect   This is the oldest (Brock, 1954; Bardach, 1959) and the most frequently used method of visually surveying fishes and macroinvertebrates. It also has been the most widely applied to natural and artificial reef surveys, probably due to its simplicity and well-defined protocol. However, its application varies considerably among users. The transect or strip transect is censused by a diver, or sometimes by submarine or ROV, moving in a direction of predetermined distance or time. Timed transects are usually employed when movie or video cameras are used, as the exposure time of the film or videotape often predetermines transect "length" (Alevizon and Brooks, 1975; Bortone *et al.*, 1986; Seaman *et al.*, 1989b). The physical distance of transects can vary considerably. Thresher and Gunn (1986) surveyed a 500 m transect while being towed from a boat. Prince *et al.* (1985) used a transect length of 180 m. Anderson *et al.* (1989) used transects as short as 5 m. Other studies have employed transects of varying lengths (Jessee *et al.*, 1985; Anderson *et al.*, 1989; DeMartini *et al.*, 1989).

   When the bottom is heterogeneous with regard to habitat it is especially important to keep the transects short. A long transect can traverse many habitats (and may include ecotonal areas that are noted to concentrate or combine faunas; Anderson *et al.*, 1989) and provide data that do not allow the variation in population parameters to be attributed to habitat association (Jones and Chase, 1975). Harmelin-Vivien *et al.* (1985) suggested using long transects for homogeneous areas and shorter transects for heterogeneous zones. Transects perpendicular to the depth contour should be avoided as habitat and species mix often change with depth (Harmelin-Vivien *et al.*, 1985).

   Divers generally swim along the transect, recording species abundance and size in a three-dimensional corridor that can vary considerably in width and height (Fig. 6.6). Brock (1954) used a transect width of 6 m; Ambrose and Swarbrick (1989) and Anderson *et al.* (1989) used a 1-m-wide transect to study recruitment. Most studies have employed a width of either 1 or 2 m (e.g., McCain and Peck, 1973; Kimmel, 1985a; Lindquist *et al.*, 1985). The rationale for narrow width is that the diver can observe smaller fishes (Harmelin-Vivien *et al.*, 1985; Sale and Sharp, 1983) and easily estimate it underwater (i.e., 1 m is about one arm length from the transect).

   Width appears to be a critical component of transect methodology. Sale and Sharp (1983) found that transects of various widths returned different population density estimates in the same survey area. Much of this difference derives from the differential detection of some species due to their size and behavior, hence observability. Harmelin-Vivien *et al.* (1985) suggested using different widths for different species (e.g., 1 m or less for smaller, cryptic species such as gobies (Gobiidae), and 3–5 m for larger species such as parrotfish (Scaridae). Kevern *et al.* (1985) used a transect width equal to one-half the estimated visibility on either side of the transect line. While this is a practical modification of the method, Sale and Sharp (1983) do not recommend using variable width transects within a study as this may make density comparisons unreliable.

Because most fishes and some macroinvertebrates are nektonic and move about considerably during a visual survey, it is important to maintain constant area and time parameters between surveys or samples within a study. Comparison of two transects of identical length but sampled for different lengths of time can produce incomparable data. Caughley *et al.* (1976) indicated that the probability of missing species on aerial transect surveys increased with the speed of the observer. Some studies have tried to maintain constant swimming speeds to reduce this problem (DeMartini *et al.*, 1989; Lincoln-Smith, 1988), but this can be difficult to do under varying current conditions.

Often a preplaced line is used as a guide for defining the census area and a reference for estimating transect length and width. Setting a transect line requires effort to place and remove the line (Kimmel, 1985b), and habitat disturbance; impact on the observability of organisms potentially can occur.

Brock (1954) developed a sampling protocol for transects that outlines criteria for the inclusion or exclusion of organisms from the count. A diver includes fishes within his/her forward view and within the zone of predetermined width. If an individual organism is nearby but does not enter the zone, it is not counted. An individual, once counted, is not recounted if it reenters the area. This may be difficult to determine under some circumstances because of the multitude of individuals and their movement. Fishes behind the diver are never counted. If one member of the school enters the area the entire school with which it is associated is included in the enumeration. Species found to be attracted to the area due to the activities of the diver-observers should not be included in the enumeration (Harmelin-Vivien *et al.*, 1985). In addition, the distance and angle of the observable field should be kept constant to maintain a consistent level of detectability for all organisms (Keast and Harker, 1977).

The vertical dimension of transects has not been consistently reported (Harmelin-Vivien *et al.*, 1985). Heights from 1.5 m (Jessee *et al.*, 1985) to the vertical limits of visibility have been proposed (Hollacher and Roberts, 1985). Since a three-dimensional area defines the living space of the aquatic community, the definition of the vertical component of the space is critical to the accurate determination of density estimates (DeMartini *et al.*, 1989).

   b. Point Count (Quadrat)   While the transect method has the observer moving through a zone to enumerate fauna, in a point count or quadrat technique the observer remains at a fixed point or place and conducts the enumeration within a prescribed area or volume. This general method is particularly useful when movement by the observer is difficult. This can occur when viewing from a submersible, a ROV, or when a video recording device is used (Smith and Tyler, 1973; Bortone *et al.*, 1986, 1989; Shinn and Wicklund, 1989).

   Bohnsack and Bannerot (1986) suggested using a random series of kicks and randomly selected compass headings to choose sample points. However, if a random method of site selection is chosen, it may be necessary to avoid ecotonal areas. Most artificial reefs, however, are not large enough to allow the random selection of observation points, thus systematic surveys are probably more appropriate. Once the sampling point is chosen, the observer determines distance to the circumference of the sample area (or lateral dimensions in the case of a quadrat; Slobodkin and Fishelson, 1974). This is facilitated by use of a weighted line of fixed length.

   Quadrat area dimensions vary from study to study. Bohnsack and Bannerot (1986) determined that in the clear waters of the Caribbean a radius of 7.5 m was effective. Bortone *et al.* (1989) used a radius of 5.6 m (area, 100 m²) where water clarity was more limiting. Stephan and Lindquist (1989) used a radius of 4 m on a shipwreck reef where visibility was generally low. Laufle and Pauley (1985) used a 6 m × 6 m square quadrat to estimate the fauna of an artificial reef. Luckhurst and Luckhurst (1977) used a smaller quadrat (3 m × 3 m) to study fish recruitment.

   A problem in selecting radius length for a point count census is that the longer the radius, the more likely one will miss observing a smaller, cryptic individual at the perimeter. This might be countered somewhat by closely inspecting the survey area after the prescribed time of observation. While this increases the probability of observing rare, cryptic, or diminutive organisms for inclusion in the total faunal list, such additions should not be included in the data set for quantitative analysis since sample time is inconsistent between surveys.

   In a point count survey the observer turns at the central point for a fixed interval of time (usually five minutes; Bohnsack and Bannerot, 1986). Zahary and Hartman (1985) used a survey time of 15 min or until no new individuals were recorded for 3–5 min. One should be aware of the need to keep the variables of time and area in any protocol constant. Bortone *et al.* (1986, 1989) determined that the amounts of time and area allotted to a method were perhaps more significant in determining the community than the type of method used. Thresher and Gunn (1986) used an "instantaneous" time interval in their point counts and concluded that the density estimates obtained were relative to amount of observation time.

   c. Species–Time Random Count   Perhaps the most innovative visual survey technique is the species–time random count method developed by Jones and Chase (1975), Thompson and Schmidt (1977), and Jones and Thompson (1978). It is referred to in later studies as the rapid visual technique (RVT) (e.g., DeMartini and Roberts, 1982) or the rapid visual count (RVC) (e.g., Kimmel, 1985b). It has the unique feature of accounting for species by merely listing them and scoring them according to the time interval in which they were first observed during the survey.

   Typically a diver swims in a random (i.e., haphazard) fashion over the survey area and lists species in the time interval in which they are initially observed. A species is given a score of 5 if it is first seen within the first 10 min of the 50-min observation period; a 4 if it is first seen in the next 10 min, and so forth. The method is based on the principle used in bird faunal surveys in which abundant species are likely to be observed first and rare species are likely to be seen last (Beals, 1960). Jones and Thompson (1978), following the recommendation of Gaufin *et al.* (1956), determined that eight, 50-min surveys would account for 93.5% of the species in highly diverse coral reef areas. Replication reduces the problem of the overweighting influence of observing rare species during the first time interval. In this example the maximum achievable score of relative abundance is 40 (e.g., score of 5 for each of eight surveys), and the least possible score is 1 (e.g., a species seen last during only one of the surveys).

   The method produces faunal lists that include more species than other methods and is particularly useful for recording rare species (Kimmel, 1985b; Bortone *et al.*, 1986, 1989). Its disadvantage is that abundance is really a score of probability of encounter and is complicated by the varied spatial distribution patterns of species in the assemblage (DeMartini and Roberts, 1982). The species–time random count method produces data, therefore, that are not comparable with data from most other faunal surveys. The other visual methods, described above, can report data in terms of density per area or volume, while the species–time random count technique cannot.

   To overcome this problem Kimmel (1985a,b) developed the visual fast count (VFC) method in which the diver records the actual number of individuals of any species seen within a preselected biotope. Expected frequencies were also used in place of the arbitrary interval scores of Jones and Thompson (1978). This modification of the species–time random count technique may prove to be a useful compromise of the best features of several techniques and could be adjusted to incorporate species size and the area surveyed.

d. Special Methods   Other visual assessment methods differ from the preceding categories and are unique to a particular study. In some studies, individuals are enumerated until they are all counted (Downing *et al.*, 1985; Alevizon *et al.*, 1985; Sale and Douglas, 1981; Moffit *et al.*, 1989; Hixon and Beets, 1989). This method has been used, usually without regard to time or area, under the premise that the area is small enough for the observer to count every organism that is the target of the survey. It has been applied most often to natural and artificial reefs that are discrete and limited in size.

Bortone *et al.* (1986) surveyed a fish fauna by using strobe-illuminated, still photographs. A color transparency was exposed at 10-m intervals in each of the four major compass headings along a 100-m transect. Although a permanent record of the habitat and fauna was provided, the temporal variability and the resolution of the photographs inhibited accurate species identification and enumeration.

Boland (1983) used two video cameras to record the fauna in stereo. Organism size and transect width could be determined from the tapes. This method, while expensive, shows promise where remotely operated visual surveys are required.

The spot-mapping technique of Thresher and Gunn (1986) requires that observers make detailed maps indicating the location of individuals relative to habitat features. Nonmoving organisms, such as corals, plants, and sponges, or sedentary, territorial species are better served by this technique. The technique requires considerable time and effort but may provide detailed information about home range and habitat associations.

e. Visual Assessment Comparisons   Comparison of results of visual census techniques has occurred because of the recognized need to rigorously examine sources of bias in the data (Barans and Bortone, 1983; Thresher and Gunn, 1986). Brock (1982) compared visual transect with rotenone survey data and concluded that visual census of fishes should be restricted to diurnal species and that transect surveys underestimated cryptic and abundant species. Christensen and Winterbottom (1981) established coefficients to make visual survey results comparable to surveys conducted with rotenone, so as to adjust the abundance of cryptic and secretive species. Sale and Douglas (1981) noted that visual surveys of fish assemblages are comparable to ichthyocide collections in restricted areas, but these methods each sampled slightly different components of the fauna.

Sanderson and Solonsky (1986) compared the species–time random count (RVT) with a strip transect (STT) and found that both methods produced a similar qualitative description of the fauna and that the RVT was more cost-effective. The STT is preferable, however, when quantitative data are required (Sanderson and Solonsky, 1986). DeMartini and Roberts (1982) found good qualitative agreement between the species–time random count method and a total count method, but the RVT overemphasized the importance of widespread, rare fishes and underestimated clumped but abundant species. Comparison of the RVC, VFC, and transect methods indicated reasonably good qualitative similarity (i.e., presence or absence) for the 80 species common to all three methods; however fewer species were encountered using the transect technique. Quantitatively, VFC results were similar to the transect method (Kimmel, 1985b).

Thresher and Gunn (1986) compared the spot mapping technique to six other methods to estimate the density of highly pelagic fishes. Transect and point count methods gave different estimates of density, but the point count estimate was less variable.

Day and night surveys using six methods (i.e., transect, point count, species–time random count, cinetransect, cineturret, and still photography) at two different reef habitats produced comparable overall qualitative (i.e., species presence or absence) faunal descriptions (Bortone et al., 1986). Methods that produced more "information" (i.e., greater numbers of species and individuals) were better at categorizing faunas in terms of their habitats. The conclusion from this analysis was that observation time and size of the survey area, and not the particular census method, were the most important aspects of any method for describing a community (Bortone et al., 1986). Another study (Bortone et al., 1989) compared transect, point count, and species–time random count techniques but standardized the data for survey time and area. Before standardization, the point count and transect methods produced statistically similar faunal parameters. The point count method provided a faunal description that was less variable and had a higher and less variable species diversity parameter for the assemblage. After standardization, however, the transect method was more efficient in sampling the number of individuals. This was probably due to the closer proximity of the observer to the faunal elements during the transect.

The use of multiple methods to make accurate and precise determinations of the density estimates of reef fish assemblages may seem redundant but will probably continue. From a technological view, it offers a way of comparing and refining the assessment methods. In addition, using multiple methods for assessments is thought by some to be the only valid way to evaluate the reliability of density estimates and to determine sources of bias in the data (Thresher and Gunn, 1986).

D. Nonvisual Methods for Studying the Sessile Community

The benthic, sessile component of many aquatic communities can be effectively sampled with various surface-tended devices such as trawls, dredges, and grab samplers (Thorson, 1957; Needham and Needham, 1962; Lebo *et al.*, 1973). A portion of the habitat is damaged or removed by these gears, however. Further, they are inefficient for assessing the irregular substrate typical of artificial habitats. Other methods have been developed.

Most studies on artificial reefs use some type of settling plate that is positioned temporarily on the reef, then removed and later examined in the laboratory (Fig. 6.10). Harriott and Fisk (1987) compared types of settlement plates in experiments on coral recruitment off Australia. The advantage of using removable sampling surfaces is that they can be experimentally manipulated to test various hypotheses (Hixon and Brostoff, 1985; Schoener and Schoener, 1981). Russ (1980), for example, isolated several plates with wire mesh to investigate the effect of grazing animals on the sessile community.

Figure 6.10 These four settling plates of PVC plastic are mounted on a concrete block for study of epibenthos on artificial habitat (from Hixon and Brostoff, 1985).

Plate size has varied among studies (e.g., 14.5 and 103 cm², Osman, 1977; 50 cm², Hixon and Brostoff, 1985; 15 cm², Fitzhardinge and Bailey-Brock, 1989). Materials used as settling plates (Fig. 6.10) include dead *Acropora* coral (Rogers *et al.*, 1984); cement blocks, terra cotta tiles, and plexiglas plates (Birkeland, 1977); red bricks (Woodhead and Jacobson, 1985); polyvinyl chloride (PVC) plastic pipes (Birkeland *et al.*, 1981); and corrugated quarry tiles (Baggett and Bright, 1985).

A suction or air-lift sampler is effective for removing members of the benthic community from the reef surface (Gale and Thompson, 1975; Gannon *et al.*, 1985; Benson, 1989; Hueckel *et al.*, 1989). Meanwhile, infauna can be sampled adjacent to the reef either by pushing a tubular device into the the bottom to obtain a core of predetermined depth (Hueckel *et al.*, 1989), or with various grab samplers. These methods may be more effective when operated by divers rather than by remote.

In the laboratory, sample analysis typically includes the number of species, number of individuals, and biomass (Bailey-Brock, 1989; Hixon and Brostoff, 1985; Osman, 1977; Fitzhardinge and Bailey-Brock, 1989). Percent cover has been estimated for sessile invertebrates from photographs of plates or by measuring the actual surface (Schoener and Schoener, 1981; Bailey-Brock, 1989). Carter *et al.* (1985) examined the relative importance of sessile organisms both on the surface of the settling plates and observed those that grew over others. Rutecki *et al.* (1985) directly measured the length of the periphyton, *in situ*, to determine the growth of plant material attached to the reef surface. The time and duration of immersion of the settling plates is critical to the understanding of the colonization–defaunation aspects of succession (Fitzhardinge and Bailey-Brock, 1989).

E. Visual Methods for Sampling the Sessile Community

Underwater field observations of organisms started as early as 1785 when Calvolini collected specimens from submarine caves in Sorento, Italy (Riedl, 1966). Since that time many survey techniques have been developed, primarily by terrestrial plant ecologists. Qualitative studies establishing species lists and zonation patterns remain popular (e.g., Goreau, 1959; Goreau and Wells, 1967; Goreau and Goreau, 1973; Scatterday, 1974; Bak, 1975; Colin, 1977; Taylor, 1978; Wheaton and Jaap, 1988; Jaap *et al.*, 1989). Quantitative benthic research began with minimal area concepts in submarine ecology (Gislen, 1930). A popular method of quantifying macroinvertebrates and plants has been through the use of quadrat sampling along transects, as developed by phytosociologists and marine ecologists (Stoddart, 1969; Scheer, 1978). Transects have been of great utility to document species of

invertebrates and algae (Hueckel and Buckley, 1989). Diver or remotely operated photographic techniques (both still and video) have been used successfully (Lundälv, 1971; Bohnsack, 1979; Weinberg, 1981; Van Dolah, 1983;
Baynes and Szmant, 1989; Bergstedt and Anderson, 1990).

Macroinvertebrates and macroalgae (i.e., those easily seen with the na-
ked eye; Hueckel and Buckley, 1989) are the life forms that are generally
surveyed using still photographs or quadrat sampling. The still photographic
technique has been used to estimate species composition, number of organisms and even percent cover and relative density of the sessile community (Bohnsack, 1979; Baynes and Szmant, 1989). Lundälv (1971) used
stereo photography for the *in situ* assessment of size and growth of sessile
organisms.

F. Assessing Abiotic Variables

Many aspects of assessing and monitoring biotic variables also apply
to the methods used to measure abiotic variables, including surface- and
SCUBA-tended methods (Loder *et al.*, 1974). Information on variables such
as date of deployment, distance from shore, and reef construction materials
can be obtained from the permit application forms that must be filed for
most reef construction projects. Bortone and Van Orman (1985a) were
able to obtain information on the abiotic aspects of the reefs they studied
from a previously prepared artificial reef atlas for Florida waters (Pybas,
1988). In this section we review some of the more widely proposed methods of measuring the most frequently studied abiotic features of artificial
habitats.

1. Measuring the Habitat

It is much easier to determine the three-dimensional size parameters of
artificial habitat before it is deployed (Moffit *et al.*, 1989). This description
includes the area and volume of each material used, as well as surface texture. This, however, is rarely done. Usually a dive team makes an *in situ*
survey to determine the amount of each type of material used (Bortone and
Van Orman, 1985a) and their surface texture. Rugosity (i.e., the total amount
of surface area available for habitation; Chandler *et al.*, 1985) can be used to
examine the ratio of living biomass to surface area (Luckhurst and Luckhurst, 1977). Such evaluations must be done consistently and based on quantitative criteria.

To be able to describe the role of reef building materials and surface
types in community succession, the history of deployment must be known.
This is often determined from permit records or published information, including newspaper accounts or navigation charts for shipwrecks (Ambrose

and Swarbrick, 1989; DeMartini *et al.*, 1989). The periodic addition of materials to reefs should be documented.

Divers often use devices as simple as a tape measure for the *in situ* measurement of artificial reefs (Bortone and Van Orman, 1985a; Alevizon *et al.*, 1985). From these measurements it is possible to calculate cross-sectional area and reef volume (Grove and Sonu, 1985).

In addition to using diver measurements (Carter *et al.*, 1985; Feigenbaum *et al.*, 1985), some studies have relied on depth recorders and positioning relative to buoys or LORAN coordinates to determine habitat configuration (Gannon *et al.*, 1985). Recently sidescan sonar has been employed to determine position, size, and extent of submerged artificial reef materials (Lukens *et al.*, 1989; Matthias, 1990).

Although overall reef configuration may be carefully planned before deployment, its actual orientation in place is often quite different because of problems during deployment. Reefs should be carefully monitored after deployment, as well, to detect movement.

2. Sediments and Substrate

The composition of bottom materials on which artificial habitats are deployed can be studied by remote, surface-tended operations. Most studies have used divers, however (Mathews, 1985). Sediments can be retrieved by hand (Bortone and Van Orman, 1985a) or with coring devices similar to those used to measure the infauna (Mathews, 1985). Baynes and Szmant (1989) determined the erosional potential and impact of water flow on the substrate by using "ciod cards."

Knowledge of sediment composition and texture is important in reef placement (Chandler *et al.*, 1985). This can be accomplished visually for coarse sand or gravel substrates (Carter *et al.*, 1985), or by laboratory techniques for smaller grain sizes (Folk, 1968; Baynes and Szmant, 1989). To determine the load bearing suitability of a site, a penetrometer (Jones, 1980) or an even more elaborate vane shear apparatus (Dill, 1965) could be used.

Charts and maps, if accurate and of adequate detail, are good sources of information on bottom topography and profile. DeMartini *et al.* (1989) and Ambrose and Swarbrick (1989) used maps and a planimeter to determine the area covered by habitat types such as kelp cover.

3. Water Column

From an engineering perspective, as discussed in Chapter 4, and from a biological point of view water currents are clearly an important variable to measure when studying artificial habitats. Lindquist and Pietrafesa (1989) used firmly anchored current meters at different depths. Diver-implaced current meters and hand-held compasses also have been used (Baynes and

Szmant, 1989). Bray (1981) used a stopwatch to measure the time that water-borne particles moved a fixed distance. Sanders *et al.* (1985) utilized a lunar phase index to quantitatively characterize this variable.

Most studies of light penetration use a Secchi disk in both a horizontal and vertical mode (Lindquist and Pietrafesa, 1989; Baynes and Szmant, 1989). For greater precision other studies have measured the amount of light penetrating to a specified depth using a measure of *in situ* irradiance both at the surface and at depth (Carter *et al.*, 1985). Although not strictly considered an abiotic feature, chlorophyll *a* is clearly an important consideration in evaluating aquatic productivity (Ardizzone *et al.*, 1989).

VI. Summary

Aquatic ecology is a rapidly changing field, and one should not expect the theory or methods presented herein to remain static. When designing an assessment or monitoring study the reader is encouraged to contact personnel at research institutions for the latest information regarding implementation of these methods. Moreover, individuals interested in conducting this type of research for the first time should solicit the aid of those with experience in the field. This concluding section is organized as a checklist of major considerations for assessment of artificial habitats.

- The purpose or main goal for constructing the reef should be determined.
- A research or study plan should be formulated to determine if the reason for constructing a reef has been met. This is best done by posing a series of questions to be answered about the reef. The questions should be designed so that collected data could reject the hypotheses or statements about the reef or its circumstances.
- Studies comparing artificial reefs before and after deployment and comparing artificial habitats to natural habitats are important to evaluate the variables associated with a reef.
- The data gathered should be of the appropriate kind and degree of accuracy to answer the questions posed.
- A list of the abiotic and biotic variables likely to help answer the questions posed should be made.
- The variables examined should be clearly defined as being either dependent or independent.
- Measured variables should be described statistically. Attempts should be made to determine how the variation in the dependent variables are related to the variation in the independent variables.

- Assessment studies should consider general problems associated with sampling the aquatic environment.
- The number of samples should be large enough to attribute the sources of variation found among the variables.
- The frequency of sampling should be more frequent than the anticipated frequency of the change in the variables suspected of influencing the reef parameters.
- Much data can be gleaned from individual specimens, especially by monitoring the changes in their life history features relative to environmental changes.
- If the rationale for building a reef is to improve fishing then it is important to design a study that is able to detect appropriate changes.
- Visual assessment and monitoring methods can provide an effective way to evaluate an artificial reef, but researchers should be aware of their limitations. Water condition has an especially significant effect on the quality of data obtained.
- Observers using visual assessments require adequate training in diving and use of the specific technique.
- Voucher specimens should be obtained whenever species identification is questionable.
- Population enumeration can be facilitated by using a search image for species that are similar in some life history features.
- For larger populations, counting groups of individuals rather than single individuals can facilitate enumerations.
- Size of organisms should be estimated relative to some known size or distance.
- Transect methods for visual assessment should be of constant length, width, and height dimensions to permit reliable comparisons of samples within a study.
- In heterogeneous environments transects should be of a shorter length. Care should be taken to make sure one transect does not include more than one type of habitat.
- Point count surveys or total faunal counts are more appropriate to areas that are limited in size or those that are heterogeneous.
- Visual techniques are easily modified to the special circumstances of each situation. Care must be taken to assure that the sampling protocol is well-defined and closely followed.
- The amount of survey time and area are critical variables to visual survey methods.
- Removable sampling plates are effective for studying the sessile flora and fauna on artificial reefs.

- Some physical features of an artificial habitat are best measured prior to deployment.
- Sediment and substrate type should be monitored as they can indicate the impact that the reef structure has on local currents and water movement.
- Those unfamiliar with sampling design and protocol should consult with specialists in these areas before attempting to design an assessment or monitoring study of artificial habitats.

References

Ahr, W. M. 1974. Geological considerations for artificial reef site locations. Pp. 31–33 *in* L. Colunga and R. B. Stone, editors. Proceedings of an International Conference on Artificial Reefs, TAMU-SG-74-103. Texas A & M University Sea Grant College Program, College Station.

Alevizon, W. S., and M. B. Brooks. 1975. The comparative structure of two western Atlantic reef-fish assemblages. Bulletin of Marine Science 25:482–490.

Alevizon, W. S., and J. C. Gorham. 1989. Effects of artificial reef development on nearby resident fishes. Bulletin of Marine Science 44:646–661.

Alevizon, W. S., J. C. Gorham, R. Richardson, and S. A. McCarthy. 1985. Use of man-made reefs to concentrate snapper (Lutjanidae) and grunts (Haemulidae) in Bahamian waters. Bulletin of Marine Science 37:3–10.

Ambrose, R. F., and S. L. Swarbrick. 1989. Comparison of fish assemblages on artificial and natural reefs off the coast of southern California. Bulletin of Marine Science 44:718–733.

Anderson, T. W., E. E. DeMartini, and D. A. Roberts. 1989. The relationship between habitat structure, body size and distribution of fishes at a temperate artificial reef. Bulletin of Marine Science 44:681–697.

Andrew, N. L., and B. D. Mapstone. 1987. Sampling and the description of spatial pattern in marine ecology. Oceanography and Marine Biology 25:39–90.

Anonymous. 1974. Artificial reefs for Texas, TAMU-SG-73-214. Texas Coastal and Marine Council, Texas A & M University Sea Grant College, College Station.

Ardizzone, G. D., M. F. Gravina, and A. Belluscio. 1989. Temporal development of epibenthic communities on artificial reefs in the central Mediterranean Sea. Bulletin of Marine Science 44:592–608.

Baggett, L. S., and T. J. Bright. 1985. Coral recruitment at the East Flower Garden Reef (Northwest Gulf of Mexico). Proceedings of the Fifth International Coral Reef Symposium 4:379–384.

Bailey-Brock, J. H. 1989. Fouling community development on an artificial reef in Hawaiian waters. Bulletin of Marine Science 44:580–591.

Bak, R. 1975. Ecological aspects of distribution of reef corals in the Netherlands Antilles. Bijdragen tot de Dierkunde 45:181–190.

Barans, C. A., and S. A. Bortone, editors. 1983. The visual assessment of fish populations in the southeastern United States: 1982 workshop, Technical Report 1 (SC-SG-TR-01-83). South Carolina Sea Grant Consortium, Charleston.

Bardach, J. E. 1959. The summer standing crop of fish on a shallow Bermuda reef. Limnology and Oceanography 4:77–85.

Baynes, T. W., and A. M. Szmant. 1989. Effect of current on the sessile benthic community structure of an artificial reef. Bulletin of Marine Science 44:545–566.

Beals, E. 1960. Forest bird communities in the Apostle Islands of Wisconsin. Wilson Bulletin 72(2):156–181.

Beets, J. 1989. Experimental evaluation of fish recruitment to combinations of fish aggregating devices and benthic artificial reefs. Bulletin of Marine Science 44:973–983.

Begon, M., J. L. Harper, and C. R. Townsend. 1986. Ecology: Individuals, populations, communities. Sinauer Associates, Sunderland, Massachusetts.

Bell, J. D., G. J. S. Craik, D. A. Pollard, and B. C. Russell. 1985. Estimating length frequency distributions of large reef fish underwater. Coral Reefs 4:41–44.

Bell, M., C. J. Moore, and S. W. Murphey. 1989. Utilization of manufactured reef structures in South Carolina's marine artificial reef program. Bulletin of Marine Science 44:818–830.

Benson, B. L. 1989. Airlift sampler: Applications for hard substrata. Bulletin of Marine Science 44:752–756.

Bergstedt, R. A., and D. R. Anderson. 1990. Evaluation of line transect sampling based on remotely censused data from underwater video. Transactions of the American Fisheries Society 119:86–91.

Bevan, W. R., A. Maier, and H. Helson. 1963. The influence of context upon the estimation of number. American Journal of Psychology 76:464–469.

Birkeland, C. 1977. The importance of rate of biotic accumulation in early successional stages of benthic communities to the survival of coral recruits. Proceedings of the Third International Coral Reef Symposium 1:15–21.

Birkeland, C., D. Rowley, and R. H. Randall. 1981. Coral recruitment patterns at Guam. Proceedings of the Fourth International Coral Reef Symposium 2:339–344.

Bohnsack, J. A. 1979. Photographic quantitative sampling studies of hard-bottom benthic communities. Bulletin of Marine Science 29:242–252.

Bohnsack, J. A. 1989. Are high densities of fishes at artificial reefs the result of habitat limitation or behavioral preference? Bulletin of Marine Science 44:631–645.

Bohnsack, J. A., and S. P. Bannerot. 1986. A stationary visual census technique for quantitatively assessing community structure of coral reef fishes. U. S. Dept. of Commerce, NOAA Technical Report NMFS 41:1–15.

Bohnsack, J. A., and D. E. Harper. 1988. Length-weight relationships of selected marine reef fishes from the southeastern United States, NOAA Technical Memorandum NMFS-SEFC. U.S. Department of Commerce, Miami, Florida.

Bohnsack, J. A., and D. L. Sutherland. 1985. Artificial reef research: A review with recommendations for future priorities. Bulletin of Marine Science 37:11–39.

Boland, G. L. 1983. A technique for determining length of free-swimming fishes using underwater stereo television. Pp. 10–11 in C.A. Barans and S.A. Bortone, editors. The visual assessment of fish populations in the southeastern United States: 1982 workshop. Technical Report 1 (SC-SG-TR-01-83). South Carolina Sea Grant Consortium, Charleston.

Bortone, S. A. 1976. Effects of a hurricane on the fish fauna at Destin, Florida. Florida Scientist 39(4):245–248.

Bortone, S. A., and D. Van Orman. 1985a. Biological survey and analysis of Florida's artificial reefs, Technical Paper 34. Florida Sea Grant College, Gainesville.

Bortone, S. A., and D. Van Orman. 1985b. Database formation and assessment of biotic and abiotic parameters associated with artificial reefs, Technical Paper 35. Florida Sea Grant College, Gainesville.

Bortone, S. A., R. W. Hastings, and J. L. Oglesby. 1986. Quantification of reef fish assemblages: A comparison of several in situ methods. Northeast Gulf Science 8:1–22.

Bortone, S. A., J. J. Kimmel, and C. M. Bundrick. 1989. A comparison of three methods for visually assessing reef fish communities: Time and area compensated. Northeast Gulf Science 10:85–96.

Bortone, S. A., R. L. Shipp, W. P. Davis, and R. D. Nester. 1988. Artificial reef development along the Atlantic coast of Guatemala. Northeast Gulf Science 10:45–48.

Bray, R. N. 1981. Influence of water movements and zooplankton densities on daily foraging movements of blacksmith, *Chromis puntipinnis*, a planktivorous fish. Fishery Bulletin 78:829–841.

Brock, R. E. 1982. A critique of the visual census method for assessing coral reef fish populations. Bulletin of Marine Science 32:269–276.

Brock, R. E. 1985. Preliminary study of the feeding habits of pelagic fish around Hawaiian fish aggregation devices or can fish aggregations devices enhance local fisheries? Bulletin of Marine Science 37:40–49.

Brock, R. E., R. M. Buckley, and R. A. Grace. 1985. An artificial reef enhancement program for nearshore Hawaiian waters. Pp. 317–336 *in* F. M. D'Itri, editor. Artificial reefs: Marine and freshwater applications. Lewis Publishers, Inc., Chelsea, Michigan.

Brock, R. E., and J. E. Norris. 1989. An analysis of the efficacy of four artificial reef designs in tropical waters. Bulletin of Marine Science 44:934–941.

Brock, V. E. 1954. A preliminary report on a method of estimating reef fish population. Journal of Wildlife Management 18(3):297–317.

Buchanon, C. C. 1974. Comparative study of the sport fishery over artificial and natural habitats off Murrells Inlet, S. C Pp. 34–38 *in* L. Colunga and R. B. Stone, editors. Proceedings of an International Conference on Artificial Reefs, TAMU-SG-74-103. Texas A & M University Sea Grant College Program, College Station.

Buckley, R. M., and G. J. Hueckel. 1985. Biological processes and ecological development on an artificial reef in Puget Sound, Washington. Bulletin of Marine Science 37:50–69.

Buckley, R. M., and G. J. Hueckel. 1989. Analysis of visual transects for fish assessment on artificial reefs. Bulletin of Marine Science 44:893–898.

Buckley, R. M., D. G. Itano, and T. W. Buckley. 1989. Fish aggregation device (FAD) enhancement of offshore fisheries in American Samoa. Bulletin of Marine Science 44:942–949.

Caddy, J. F., and G. P. Bozigos. 1985. Guidelines for statistical monitoring, FAO Fisheries Technical Paper 257:1–86. United Nations Food and Agricultural Organization, Rome.

Cairns, J., W. Younge, and R. Kaesler. 1976. Qualitative differences in protozoan colonization of artificial substrates. Hydrobiologica 5:26–46.

Campos, J. A., and C. Gamboa. 1989. An artificial tire-reef in a tropical marine system: A management tool. Bulletin of Marine Science 44:757–766.

Carter, J. W., A. L. Carpenter, M. S. Foster, and W. N. Jessee. 1985. Benthic succession on an artificial reef designed to support a kelp-reef community. Bulletin of Marine Science 37:86–113.

Caughley, G., R. Sinclair, and D. Scott-Kemmis. 1976. Experiments in aerial surveys of faunal populations. Journal of Wildlife Management 40:290–300.

Chandler, C. R., R. M. Sanders, Jr., and A. M. Landry, Jr. 1985. Effects of three substrate variables on two artificial reef communities. Bulletin of Marine Science 37:129–142.

Christensen, M. S., and R. Winterbottom. 1981. A correction factor for, and its application to, visual census of littoral fish. Suid-Afrikaanse Tydskr; vir Diergeneeskunde. 16(2):73–79.

Cole, G. A. 1983. Textbook of limnology. C. V. Mosby Company, St. Louis, Missouri.

Colin, P. L. 1977. The reefs of Cocos-Keeling Atoll, eastern Indian Ocean. Proceedings of the Third International Coral Reef Symposium 1:63–68.

Collette, B. B., and S. A. Earle. 1972. The results of the Tektite program: Ecology of coral reef fishes. Los Angeles County Museum, Science Bulletin 14:1–180.

Colton, D. E., and W. S. Alevizon. 1981. Diurnal variability in a fish assemblage of a Bahamian coral reef. Environmental Biology of Fishes 6:341–435.

Connell, J. H. 1978. Diversity in tropical rainforests and coral reefs. Science 199:1302–1310.

Corbetta, M., F. M Miezin, S. Dobmeyer, G. L. Shulman, and S. E. Petersen. 1990. Attentional modulation of neural processing of shape, color, and velocity in humans. Science 248:1556–1559.

Crumpton, J. E., and R. L. Wilbur. 1974. Florida's fish attractor program. Pp. 39–46 in L. Colunga and R. B. Stone, editors. Proceedings of an International Conference on Artificial Reefs, TAMU-SG-74-103. Texas A & M University Sea Grant College Program, College Station.

Dale, G. 1978. Money-in-the-bank: A model for coral reef fish coexistence. Environmental Biology of Fishes 3:103–108.

Davis, G. E. 1985. Artificial structures to mitigate marina construction impacts on spiny lobster, *Panulirus argus*. Bulletin of Marine Science 37:151–156.

Dawson, C. E. 1965. Length-weight relationships of some Gulf of Mexico fishes. Transactions of the American Fisheries Society 94:279–280.

DeMartini, E. E., and D. Roberts. 1982. An empirical test of biases in the rapid visual technique for census of reef fish assemblages. Marine Biology 70:129–134.

DeMartini, E. E., D. A. Roberts, and T. W. Anderson. 1989. Contrasting patterns of fish density and abundance at an artificial rock reef and a cobble-bottom kelp forest. Bulletin of Marine Science 44:881–892.

Dennis, G. D., and T. J. Bright. 1988. Reef fish assemblages on hard banks in the northwestern Gulf of Mexico. Bulletin of Marine Science 43:280–307.

Dill, R. F. 1965. A diver-held vane-shear apparatus. Marine Geology 3:323–327.

Downing, N., R. A. Tubb, C. R. El-Zahr, and R. E. McClure. 1985. Artificial reefs in Kuwait, northern Arabian Gulf. Bulletin of Marine Science 37:157–178.

Ebeling, A. W., R. J. Larson, W. S. Alevizon, and R. N. Bray. 1980. Annual variability of reef fish assemblages in kelp forests off Santa Barbara, California. Fishery Bulletin 78:361–377.

Emmel, T. C. 1976. Population biology. Harper & Row, New York.

Enns, J. T., and R. A. Rensink. 1990. Influence of scene-based properties on visual search. Science 247:721–723.

Epperly, S. P. 1983. Comparison of underwater visual techniques for reef fish censusing. Pp. 15–16 in C. A. Barans and S. A. Bortone, editors. The visual assessment of fish populations in the southeastern United States: 1982 workshop, Technical Report 1 (SC-SG-TR-01-83). South Carolina Sea Grant Consortium, Charleston.

Fabi, G., L. Fiorentini, and S. Giannini. 1989. Experimental shellfish culture on an artificial reef in the Adriatic Sea. Bulletin of Marine Science 44:923–933.

Fast, D. E., and F. A. Pagan. 1974. Comparison of fishes and biomass by trophic levels on artificial and natural reefs off southwestern Puerto Rico. P. 58 in L. Colunga and R. B. Stone, editors. Proceedings of an International Conference on Artificial Reefs, TAMU-SG-74-103. Texas A & M University Sea Grant College Program, College Station.

Feigenbaum, D., C. H. Blair, M. Bell, J. R. Martin, and M. G. Kelly. 1985. Virginia's artificial reef study: Description and results of Year I. Bulletin of Marine Science 37:179–188.

Feigenbaum, D., M. Bushing, J. Woodward, and A. Freidlander. 1989a. Artificial reefs in Chesapeake Bay and nearby coastal waters. Bulletin of Marine Science 44:734–742.

Feigenbaum, D., A. Freidlander, and M. Bushing. 1989b. Determination of the feasibility of fish attracting devices for enhancing fisheries in Puerto Rico. Bulletin of Marine Science 44:950–959.

Fitzhardinge, R. C., and J. H. Bailey-Brock. 1989. Colonization of artificial reef materials by corals and other sessile organisms. Bulletin of Marine Science 44:567–579.

Folk, R. L. 1968. Petrology of sedimentary rocks. Hemphill's, Austin, Texas.

Gale, W. F., and J. D. Thompson. 1975. A suction sampler for quantitatively sampling benthos of rock substrates in rivers. Transactions of the American Fisheries Society 104:398–405.

Galzin, R. 1985. Ecologie des poissons récifaux de Polynésie française: Variations spatio-temporelles des peuplements. Dynamique de populations de trois espèces dominantes des lagons nord de Moores. Evaluations de la production ichtyolgique d'un secteur récifolagonaire. Thèsis de Doctorat ès Sciences, Université de Sciences et Techniques du Languedoc.

Gannon, J. E., R. J. Danehy, J. W. Anderson, G. Merritt, and A. P. Bader. 1985. The ecology of natural shoals in Lake Ontario and their importance to artificial reef development. Pp. 113–139 in F. M. D'Itri, editor. Artificial reefs: Marine and freshwater applications. Lewis Publishers, Inc., Chelsea, Michigan.

Gauch, H. G., Jr. 1982. Multivariate analysis in community ecology. Cambridge University Press, Cambridge, England.

Gaufin, A. R., E. K. Hanis, and H. J. Walter. 1956. A statistical evaluation of stream bottom sampling data obtained from three standard samples. Ecology 37:643–648.

Ginsburg, A. P., and M. W. Conner. 1983. Comparison of three methods for rapid determination of threshold contrast sensitivity. Investigative Ophthalmology & Visual Science 24:798–802.

Gislén, T. 1930. Epibiosis of the Gulmar Fjord II. Marine sociology. Skriftser. Kungl. Svenska Vetenskapsakad. 4:1–380.

Gladfelter, W. B., J. C. Ogden, and E. H. Gladfelter. 1980. Similarity and diversity among coral reef fish communities: A comparison between tropical Western Atlantic (Virgin Islands) and tropical central Pacific (Marshall Islands) patch reefs. Ecology 61:1156–1168.

Glynn, P. W. 1973. Ecology of Caribbean coral reefs. The Porites reef-flat biotope. Part II. Plankton communities with evidence for depletion. Marine Biology 22:1–21.

Goldman, B., and F. H. Talbot. 1976. Aspects of the ecology of coral reef fishes. Pp. 125–154 in O. H. Jones and R. Endean, editors. Biology and ecology of coral reefs, Volume 3. Academic Press, New York.

Goreau, T. F. 1959. The ecology of Jamaican coral reefs. I. Species composition and zonation. Ecology 40:67–90.

Goreau, T. F., and N. I. Goreau. 1973. The ecology of Jamaican coral reefs. II. Geomorphology, zonation, and sedimentary phases. Bulletin of Marine Science 23:399–464.

Goreau, T. F., and J. W. Wells. 1967. The shallow water Scleratinia of Jamaica: Revised list of species and their vertical range. Bulletin of Marine Science 17:447–453.

Gorham, J. C., and W. S. Alevizon. 1989. Habitat complexity and the abundance of juvenile fishes residing on small scale artificial reefs. Bulletin of Marine Science 44:662–665.

Gotshall, D. W. 1964. Increasing tagged rockfish (genus Sebastodes) survival by deflating the swim bladder. California Fish and Game 50:253–260.

Green, R. H. 1979. Sampling design and statistical methods for environmental biologists. Wiley, New York.

Greene, L. E., and W. S. Alevizon. 1989. Comparative accuracies of visual assessment methods for coral reef fishes. Bulletin of Marine Science 44:899–912.

Grove, R. S., and C. J. Sonu. 1983. Review of Japanese fishing reef technology, Report 83-RD-137. Southern California Edison Company, Rosemead, California.

Grove, R. S., and C. J. Sonu. 1985. Fishing reef planning in Japan. Pp. 187–251 in F. M. D'Itri, editor. Artificial reefs: Marine and freshwater applications. Lewis Publishers, Inc., Chelsea, Michigan.

Gulland, J. A. 1983. Fish stock assessment: A manual of basic methods, Volume 1. FAO/Wiley Series on Food and Agriculture, Wiley, Chichester, England.

Harmelin-Vivien, M. L., J. G. Harmelin, C. Chauvet, C. Duval, R. Galzin, P. Lejeune, G. Barnabé, F. Blanc, R. Chevalier, J. Duclerc, and G. Lasserre. 1985. Evaluation visuelle des peuplements et populations de poissons: Méthodes et problèmes. Revue d' Ecologie (Terre Vie) 40:467–539.

Harriott, V. J., and D. A. Fisk. 1987. A comparison of settlement plate types for experiments on the recruitment of scleratinian corals. Marine Ecology: Progress Series 37:201–208.

Hastings, R. W. 1979. The origin and seasonality of the fish fauna on a new jetty in the northeastern Gulf of Mexico. Florida State Museum, Biological Sciences 24(1):1–124.

Hastings, R. W., L. H. Ogren, and M. T. Mabry. 1976. Observations on the fish fauna on a new jetty in the northeastern Gulf of Mexico. Fishery Bulletin 74:387–401.

Helfman, G. S. 1978. Patterns of community structure in fishes: Summary and overview. Environmental Biology of Fishes 3:129–148.

Helfman, G. S. 1983. Underwater methods. Pp. 349–369 in L. A. Nielsen and D. L. Johnson, editors. Fisheries techniques. American Fisheries Society, Bethesda, Maryland.

Helvey, M., and R. W. Smith. 1985. Influence of habitat structure on the fish assemblages associated with two cooling-water intake structures in southern California. Bulletin of Marine Science 37:189–199.

Hixon, M. A. 1980. Competitive interaction between California reef fishes of the genus Embiotoca. Ecology 61:918–931.

Hixon, M. A., and J. P. Beets. 1989. Shelter characteristics and Caribbean fish assemblages: Experiments with artificial reefs. Bulletin of Marine Science 44:666–680.

Hixon, M. A., and W. N. Brostoff. 1985. Substrate characteristics, fish grazing, and epibenthic reef assemblages off Hawaii. Bulletin of Marine Science 37:200–213.

Hobson, E. S. 1965. Diurnal-nocturnal activity of some inshore fishes in the Gulf of California. Copeia 1965(3):291–302.

Hollacher, L. E., and D. A. Roberts. 1985. Differential utilization of space and foods by the inshore rockfishes (Scorpaenidae: Sebastes) of Carmel Bay, California. Environmental Biology of Fishes 12:91–110.

Hueckel, G. J., and R. M. Buckley. 1989. Predicting fish species on artificial reefs using indicator biota from natural reefs. Bulletin of Marine Science 44:873–880.

Hueckel, G. J., R. M. Buckley, and B. L. Benson. 1989. Mitigating rocky habitat loss using artificial reefs. Bulletin of Marine Science 44:913–922.

Jaap, W. C., W. G. Lyons, P. Dustan, and J. C. Halas. 1989. Stony coral (Scleractinea and Milleporina) community structure at Bird Key Reef, Ft. Jefferson National Monument, Dry Tortugas, Florida, Publication 46. Florida Marine Research, St. Petersburg.

Jessee, W. N., A. L. Carpenter, and J. W. Carter. 1985. Distribution patterns and density estimates of fishes on a southern California artificial reef with comparisons to natural kelp-reef habitats. Bulletin of Marine Science 37:214–225.

Jones, C. P. 1980. Engineering aspects of artificial reef failures. Notes for Florida Sea Grant's artificial reef research diver training program, NEMAP Fact Sheet 5. Florida Cooperative Extension Service, University of Florida, Gainesville.

Jones, R. S., and J. A. Chase. 1975. Community structure and distribution of fishes in an enclosed high island lagoon in Guam. Micronesica 11(1):127–148.

Jones, R. S., and M. J. Thompson. 1978. Comparison of Florida reef fish assemblages using a rapid visual technique. Bulletin of Marine Science 28:159–172.

Keast, A., and J. Harker. 1977. Strip counts as a means of determining densities and habitat utilization patterns in lake fishes. Environmental Biology of Fishes 1:181–188.

Kevern, N. R., W. E. Biener, S. R. VanDerLann, and S. D. Cornelius. 1985. Preliminary evaluation of an artificial reef as a fishery management strategy in Lake Michigan. Pp. 443–458 *in* F. M. D'Itri, editor. Artificial reefs: Marine and freshwater applications. Lewis Publishers, Inc., Chelsea, Michigan.

Kimmel, J. J. 1985a. A characterization of Puerto Rican fish assemblages. Ph.D. Dissertation, University of Puerto Rico, Mayaquez.

Kimmel, J. J. 1985b. A new species-time method for visual assessment of fishes and its comparison with established methods. Environmental Biology of Fishes 12:23–32.

Kingsford, M. J. 1989. Distribution of planktivorous reef fish along the coast of northeastern New Zealand. Marine Ecology: Progress Series 54:13–24.

Klima, E. F., and D. A. Wickham. 1971. Attraction of coastal pelagic fishes with artificial structures. Transactions of the American Fisheries Society 100:86–99.

Lang, M. A., and R. W. Hamilton, editors. 1989. Proceedings of the American Academy of Underwater Science, USCSG-TR-01-892. Diver Computer Workshop, University of California Sea Grant Program, San Diego.

Larson, R. L., and E. E. DeMartini. 1984. Abundance and vertical distribution of fishes in a cobble-bottom kelp forest off San Onofre, California. Fishery Bulletin 82:37–53.

Laufle, J. C., and G. B. Pauley. 1985. Fish colonization and materials comparisons on a Puget Sound artificial reef. Bulletin of Marine Science 37:227–243.

Lebo, L., R. Garton, P. A. Lewis, K. Mackenthun, W. T. Mason, R. Nadeau, D. Phelps, R. Schneider, and R. Sinclair. 1973. Macroinvertebrates. Pp. 1–38 *in* C. I. Weber, editor. Biological field and laboratory methods for measuring the quality of surface waters and effluents, Environmental Monitoring Series, Program Element 1BA027, EPA-670/4-73-001. U. S. Environmental Protection Agency, Washington, D. C.

Lincoln-Smith, M. P. 1988. Effects of observer swimming speed on sample counts of temperate rocky reef fish assemblages. Marine Ecology: Progress Series 43:223–231.

Lindquist, D. G., and L. J. Pietrafesa. 1989. Current vortices and fish aggregations: The current field and associated fishes around a tugboat wreck in Onslow Bay, North Carolina. Bulletin of Marine Science 44:533–544.

Lindquist, D. G., M. V. Ogburn, W. B. Stanley, H. L. Troutman, and S. M. Pereira. 1985. Fish utilization patterns on temperate rubble-mound jetties in North Carolina. Bulletin of Marine Science 37:244–251.

Liston, C. R., D. C. Brazo, J. R. Bohr, and J. A. Gulvas. 1985. Abundance and composition of Lake Michigan fishes near rock jetties and a breakwater, with comparisons to fishes in nearby natural habitats. Pp. 491–514 *in* F. M. D'Itri, editor. Artificial reefs: Marine and freshwater applications. Lewis Publishers, Inc., Chelsea, Michigan.

Loder, T. C., G. T. Rowe, and C. H. Clifford. 1974. Experiments using baled urban refuse as artificial reef material. Pp. 56–59 *in* L. Colunga and R. B. Stone, editors. Proceedings of an International Conference on Artificial Reefs, TAMU-SG-74-103. Texas A & M University Sea Grant College Program, College Station.

Luckhurst, B. E., and L. Luckhurst. 1977. Recruitment patterns of coral reef fishes on the fringing reef of Curaao, Netherlands Antilles. Canadian Journal of Zoology 55:681–689.

Luckhurst, B. E., and L. Luckhurst. 1978. Analysis of the influence of substrate variables on coral reef communities. Marine Biology 49:317–323.

Ludwig, J. A., and J. F. Reynolds. 1988. Statistical ecology. Wiley, New York.

Lukens, R. R. 1981. Ichthyofaunal colonization of a new artificial reef in the northern Gulf of Mexico. Gulf Research Reports 7:41–46.

Lukens, R. R., J. D. Cirino, J. A. Ballard, and G. Geddes. 1989. Two methods of monitoring and assessment of artificial reef materials, Special Report 2-WB. Gulf States Marine Fisheries Commission, Ocean Springs, Mississippi.

Lundälv, T. 1971. Quantitative studies on rocky-bottom biocenoses by underwater photogrammetry. Thalassia Jugoslavia 7(1):201–208.

Mathews, H. 1985. Physical and geological aspects of artificial reef sit selections. Pp. 141–148 *in* F. M. D'Itri, editor. Artificial reefs: Marine and freshwater applications. Lewis Publishers, Inc., Chelsea, Michigan.

Matthews, K. R. 1985. Species similarity and movements of fishes on natural and artificial reefs in southern California. Bulletin of Marine Science 37:252–270.

Matthias, P. 1990. Aircraft wreck surprise find during sonar tests. Sea Technology 31(3):64–65.

McCain, J. C., and J. M. Peck. 1973. The effects of a Hawaiian power plant on the distribution and abundance of reef fishes, Sea Grant Advisory Report UNIHI-SEA, Grant-AR-73-03. University of Hawaii, Honolulu.

McCormick, M. I., and J. H. Choat. 1987. Estimated total abundance of large temperate-reef fish using visual strip-transects. Marine Biology 96:469–478.

McKaye, K. R. 1977. Competition for breeding sites between the cichlid fishes of Lake Tiloá, Nicaragua. Ecology 58:291–302.

Meier, M. H., R. Buckley, and J. J. Polovina. 1989. A debate on responsible artificial reef development. Bulletin of Marine Science 44:1051–1077.

Miller, R. J., and W. Hunte. 1987. Effective area fished by antillean fish traps. Bulletin of Marine Science 40:484–493.

Milon, J. W. 1989a. Economic evaluation of artificial habitat for fisheries: Progress and challenges. Bulletin of Marine Science 44:831–843.

Milon, J. W. 1989b. Artificial marine habitat characteristics and participation behavior by sport anglers and divers. Bulletin of Marine Science 44:853–862.

Moffit, R. B., F. A. Parrish, and J. L. Polovina. 1989. Community structure, biomass and productivity of deepwater artificial reefs in Hawaii. Bulletin of Marine Science 44:616–630.

Molles, M. C., Jr. 1978. Fish species diversity on model and natural reef patches: Experimental insular biogeography. Ecological Monographs 48:289–305.

Moring, J. R., M. T. Negus, R. D. McCullough, and S. W. Herke. 1989. Large concentrations of submerged pulpwood logs as fish attraction structures in a reservoir. Bulletin of Marine Science 44:609–615.

Morrison, T. R. 1980. A review of diagnostic visual acuity, NAMRL Monograph 28. Naval Aerospace Medical Research Laboratory, Pensacola, Florida.

Myatt, D. O., E. N. Myatt, and W. K. Figley. 1989. New Jersey tire reef stability study. Bulletin of Marine Science 44:807–817.

Nakamura, M. 1985. Evolution of artificial fishing reef concepts in Japan. Bulletin of Marine Science 37:271–278.

National Oceanic and Atmospheric Administration (NOAA). 1975. The NOAA diving manual. Manned Undersea Science and Technology Office, U. S. Government Printing Office, Washington, D. C.

Needham, J. G., and P. R. Needham. 1962. A guide to the study of fresh-water biology. Holden-Day, San Francisco, California.

Nybakken, J. W. 1982. Marine biology: An ecological approach. Harper & Row, New York.

Osman, R. W. 1977. The establishment and development of a marine epifaunal community. Ecological Monographs 47:37–63.

Paine, R. T., and S. A. Levin. 1981. Intertidal landscapes: Disturbance and the dynamics of pattern. Ecological Monographs 51:145–178.

Patton, M. L., R. S. Grove, and R. F. Harman. 1985. What do natural reefs tell us about designing artificial reefs in southern California? Bulletin of Marine Science 37:279–298.

Pennak, R. W. 1964. Dictionary of zoology. Ronald Press, New York.

Pielou, E. C. 1966. The measurement of diversity in different types of biological collections. Journal of Theoretical Biology 13:131–144.

Pitcher, T. J., and P. J.B. Hart. 1982. Fisheries ecology. AVI Publishing Company, Westport, Connecticut.

Polovina, J. J., and I. Sakai. 1989. Impacts of artificial reefs on fishery production in Shimamaki, Japan. Bulletin of Marine Science 44:997–1003.

Poole, R. W. 1974. An introduction to quantitative ecology. McGraw-Hill, New York.

Prince, E. D., and P. Brouha. 1974. Progress of the Smith Mountain reservoir artificial reef project. Pp. 68–72 in L. Colunga and R. B. Stone, editors. Proceedings of an International Conference on Artificial Reefs, TAMU-SG-74-103. Texas A & M University Sea Grant College Program, College Station.

Prince, E. D., O. E. Maughan, and P. Brouha. 1985. Summary and update of the Smith Mountain Lake artificial reef project. Pp. 401–430 in F. M. D'Itri, editor. Artificial reefs: Marine and freshwater applications. Lewis Publishers, Inc., Chelsea, Michigan.

Pybas, D. W. 1988. Atlas of artificial reefs in Florida, Extension Bulletin SGEB-13. Florida Sea Grant, Gainesville.

Reese, E. S. 1975. A comparative field study of the social behavior and related ecology of reef fishes of the family Chaetodontidae. Zeitschrift fuer Tierpsychologie 37:37–61.

Relini, G., and L. O. Relini. 1989. Artificial reefs in the Ligurian Sea (northwestern Mediterranean): Aims and results. Bulletin of Marine Science 44:743–751.

Ricker, W. E., editor. 1968. Methods for assessment of fish production in freshwaters, IBP Handbook 3. Blackwell Scientific Publications, Oxford, England.

Riedl, R. 1966. Biologie der Meeresholen: Topographie, faunistik and okologie eines unterseeischen lebensraumes. Parey, Hamburg.

Robson, D. S., and H. A. Regier. 1968. Estimation of population number and mortality rates. Pp. 124–158 in W. E. Ricker, editor. Methods for assessment of fish production in freshwaters, IBP Handbook 3. Blackwell Scientific Publications, Oxford, England.

Roedel, P. M. 1975. A summary and critique of the symposium on optimum yield. Pp. 78–89 in Optimum sustainable yield as a concept in fisheries management, Special Publication 9. American Fisheries Society, Bethesda, Maryland.

Rogers, C. S., H. C. Fitz, III, M. Gilnacks, J. Beets, and J. Hardin. 1984. Scleratinian coral recruitment patterns at Salt River Canyon, St. Croix, U. S. Virgin Islands. Coral Reefs 3:69–76.

Rountree, R. A. 1989. Association of fishes with fish aggregation devices: Effects of structure size on fish abundance. Bulletin of Marine Science 44:960–972.

Russ, G. R. 1980. Effects of predation by fishes, competition, and structural complexity of the substratum on the establishment of a marine epifaunal community. Journal of Experimental Biology and Ecology 42:55–69.

Russ, G. R. 1984. Effects of protective management on coral reef fishes in the Philippines. ICLARM Newsletter, October:12–13.

Russell, B. C., F. H. Talbot, and S. Domm. 1974. Patterns of colonization of artificial reefs by coral reef fishes. Proceedings of the Second International Coral Reef Symposium 1:207–215.

Russell, B. C., F. H. Talbot, G. R.V. Andersen, and B. Goldman. 1978. Pp. 329–345 in D. R. Stoddard and R. E. Johannes, editors. Coral reefs: Research methods, Monograph on Oceanographic Methodology 5. UNESCO, Paris.

Rutecki, T. L., J. A. Dorr, and D. J. Jude. 1985. Preliminary analysis of colonization and succession of selected algae, invertebrates, and fish on two artificial reefs in inshore southeastern Lake Michigan. Pp. 459–489 in F. M. D'Itri, editor. Artificial reefs: Marine and freshwater applications. Lewis Publishers, Inc., Chelsea, Michigan.

Sale, P. F. 1978. Coexistence of coral reef fishes—a lottery for living space. Environmental Biology of Fishes 3:85–102.

Sale, P. F. 1980. Assemblages of fish on patch reefs—predictable or unpredictable. Environmental Biology of Fishes 5:243–249.

Sale, P. F., and W. A. Douglas. 1981. Precision and accuracy of visual census techniques for fish assemblages on coral patch reefs. Environmental Biology of Fishes 6:333–339.

Sale, P. F., and R. Dybdahl. 1975. Determinants of community structure for coral reef fishes in an experimental habitat. Ecology 56:1334–1355.

Sale, P. F., and B. J. Sharp. 1983. Correction for bias in visual transect censuses of coral reef fishes. Coral Reefs 2:37–42.

Samples, K. C. 1989. Assessing recreational and commercial conflicts over artificial fishery habitat use: Theory and practice. Bulletin of Marine Science 44:844–852.

Sanders, R. M., Jr., C. R. Chandler, and A. M. Landry, Jr. 1985. Hydrological, diel and lunar factors affecting fishes on artificial reefs off Panama City, Florida. Bulletin of Marine Science 37:318–327.

Sanderson, S. L., and A. C. Solonsky. 1986. Comparison of rapid visual and a strip transect technique for censusing reef fish assemblages. Bulletin of Marine Science 39:119–129.

Sato, O. 1985. Scientific rationales for fishing reef design. Bulletin of Marine Science 37:329–335.

Scatterday, J. W. 1974. Reefs and associated assemblages off Bonaire, Netherlands Antilles, and their bearing on Pleistocene and recent reef models. Proceedings of the Second International Coral Reef Symposium 2:88–106.

Scheer, G. 1978. Application of phytosociologic methods. Pp. 175–196 in D. R. Stoddart and R. E. Johannes, editors. Coral reefs: Research methods, Monographs on Oceanographic Methodology 5. UNESCO, Paris.

Schoener, A., and T. W. Schoener. 1981. The dynamics of the species-area relationship in marine fouling systems: 1. Biological correlates of changes in the species-area slope. American Naturalist 118:339–360.

Seaman, W., Jr., R. M. Buckley, and J. J. Polovina. 1989a. Advances in knowledge and priorities for research, technology and management related to artificial aquatic habitats. Bulletin of Marine Science 44:527–532.

Seaman, W., Jr., W. J. Lindberg, C. R. Gilbert, and T. K. Frazer. 1989b. Fish habitat provided by obsolete petroleum platforms off southern Florida. Bulletin of Marine Science 44:1014–1022.

Sheehy, D. J. 1985. New approaches in artificial reef design and applications. Pp. 253–263 in F. M. D'Itri, editor. Artificial reefs: Marine and freshwater applications. Lewis Publishers, Inc., Chelsea, Michigan.

Shinn, E. A., and R. I. Wicklund. 1989. Artificial reef observations from a manned submersible off southeast Florida. Bulletin of Marine Science 44:1041–1050.

Shipp, R. L., W. A. Tyler, Jr., and R. S. Jones. 1986. Point count censusing from a submersible to estimate fish abundance over large areas. Northeast Gulf Science 8:83–89.

Slobodkin, L. B., and L. Fishelson. 1974. The effect of the cleaner-fish Labroides dimidiatus on the point diversity of fishes on the reef front at Eilat. American Naturalist 108:369–376.

Smith, C. L. 1973. Small rotenone stations: A tool for studying coral reef fish communities. American Museum Novitates 2512:1–21.

Smith, C. L., and J. C. Tyler. 1972. Space resource sharing in a coral reef community. Pp. 125–170 in B. B. Collette and S. A. Earle, editors. Results of the Tektite Program: Ecology of coral reef fishes, Science Bulletin. Natural History Museum of Los Angeles, Los Angeles, California.

Smith, C. L., and J. C. Tyler. 1973. Population ecology of a Bahamian suprabenthic shore fish assemblage. American Museum Novitates 2528:1–38.

Smith, C. L., and J. C. Tyler. 1975. Succession and stability in fish communities of dome-shaped patch reefs in the West Indies. American Museum Novitates 2572:1–18.

Smith, G. B. 1975. Red tide and its impact on certain reef communities in the mid-eastern Gulf of Mexico. Environmental Letters 9(2):141–152.

Smith, G. B., H. M. Austin, S. A. Bortone, R. W. Hastings, and L. H. Ogren. 1975. Fishes of the Florida Middle Ground with comments on ecology and zoogeography, Publication 9. Florida Marine Research, St. Petersburg.

Solonsky, A. C. 1985. Fish collection and the effect of fishing activities on two artificial reefs in Monterey Bay, California. Bulletin of Marine Science 37:336–347.

Somerton, D. A., B. S. Kihkaway, and C. D. Wilson. 1988. Hook times to measure the capture time of individual fish. Marine Fisheries Review 50:1–5.

Spanier, E., M. Tom, and S. Pisanty. 1985. Enhancement of fish recruitment by artificial enrichment of man-made reefs in the southeastern Mediterranean. Bulletin of Marine Science 37:356–363.

Stanley, D. R., and C. A. Wilson. 1989. Utilization of offshore platforms by recreational fishermen and SCUBA divers off the Louisiana coast. Bulletin of Marine Science 44:767–776.

Stanton, G., D. Wilber, and A. Murray. 1985. Annotated bibliography of artificial reef research, Report 74. Florida Sea Grant College Program, Gainesville.

Starck, W. A., II, and W. P. Davis. 1966. Night habits of fishes of Alligator Reef, Florida. Ichthyologica 28:313–356.

Stephan, C. D., and D. G. Lindquist. 1989. A comparative analysis of the fish assemblages associated with old and new shipwrecks and fish aggregating devices in Onslow Bay, North Carolina. Bulletin of Marine Science 44:698–717.

Stewart-Oaten, A., W. W. Murdoch, and K. R. Parker. 1986. Environmental impact assessment: "Pseudoreplication" in time? Ecology 67:929–940.

Stoddart, D. R. 1969. Reef studies at Adder Atoll, Marshall Islands. Atoll Research Bulletin 116:1–22.

Stott, B. 1970. Some factors affecting the catching power of unbaited fish traps. Journal of Fish Biology 2:15–22.

Sutherland, J. P. 1974. Multiple stable points in natural communities. American Naturalist 108:859–873.

Sutherland, J. P., and R. H. Karlson. 1977. Development and stability of the fouling community at Beaufort, North Carolina. Ecological Monographs 47:425–446.

Tait, R. V. 1981. Elements of marine ecology. Butterworths, London.

Talbot, F. H., B. C. Russell, and G. R.V. Anderson. 1978. Coral reef fish communities, unstable, high diversity systems. Ecological Monographs 48:425–440.

Taylor, J. D. 1978. Zonation of rocky intertidal surfaces. Pp. 139–148 in D. R. Stoddart and R. E. Johannes, editors. Coral reefs: Research methods, Monographs on Oceanographic Methodology 5. UNESCO, Paris.

Theodorakis, C. W. 1989. Size segregation and the effects of oddity on predation risk in minnow schools. Animal Behaviour 38:496–502.

Thompson, M. J., and T. W. Schmidt. 1977. Validation of the species/time random count technique for sampling fish assemblages. Proceedings of the Third International Coral Reef Symposium 1:283–288.

Thorne, R. E., J. B. Hedgepeth, and J. Campos. 1989. Hydroacoustic observations of fish abundance and behavior around an artificial reef in Costa Rica. Bulletin of Marine Science 44:1058–1064.

Thorson, G. 1957. Sampling the benthos. Pp. 61–86 in J. W. Hedgpeth, editor. Treatise on marine ecology and paleoecology. Volume 1. Ecology. Geological Society of America, New York.

Thresher, R. E., and J. S. Gunn. 1986. Comparative analysis of visual census techniques for highly mobile, reef-associated piscivores (Carangidae). Environmental Biology of Fishes 17:93–116.

Turner, C. H., E. E. Ebert, and R. R. Given. 1969. Man-made reef ecology. California Department of Fish and Game Fishery Bulletin 146:1–221.

Van Dolah, R. F. 1983. Remote assessment techniques for large benthic invertebrates. Pp. 12–13 in C. A. Barans and S. A. Bortone, editors. The visual assessment of fish populations in the southeastern United States: 1982 workshop, Technical Report 1 (SC-SG-TR-01-83). South Carolina Sea Grant Consortium, Charleston.

Weinberg, S. 1981. A comparison of coral reef survey methods. Bijdragen tot de Dierkunde 51(2):199–218.

Wheaton, J. L., and W. C. Jaap. 1988. Corals and other prominent benthic Cnidaria of Looe Key National Marine Sanctuary, Publication 43. Florida Marine Research, St. Petersburg.

Wickham, D. A., and G. M. Russell. 1974. An evaluation of mid-water artificial structures for attracting coastal pelagic fishes. Fishery Bulletin 72:181–191.

Williams, R., and D. Briton. 1986. Speech recognition as a means of enumeration in the analysis of biological samples. Marine Biology 92:595–598.

Witzig, J. F. 1983. Statistical analysis of community data. Pp. 32–33 in C. A. Barans and S. A. Bortone, editors. The visual assessment of fish populations in the southeastern United States: 1982 workshop, Technical Report 1 (SC-SG-TR-01-83). South Carolina Sea Grant Consortium, Charleston.

Woodhead, P. M. J., and M. E. Jacobson. 1985. Epifaunal settlement, the process of community development and succession over two years on an artificial reef in the New York Bight. Bulletin of Marine Science 37:364–376.

Woodhead, P. M. J., J. H. Parker, and I. W. Duedall. 1985. The use of by-products from coal combustion for artificial reef construction. Pp. 265–292 in F. M. D'Itri, editor. Artificial reefs: Marine and freshwater applications. Lewis Publishers, Inc., Chelsea, Michigan.

Workman, I. K., A. M. Landry, Jr., J. W. Watson, Jr., and J. W. Blackwell. 1985. A midwater fish attraction device study conducted from Hydrolab. Bulletin of Marine Science 37:377–386.

Zahary, R. G., and M. J. Hartman. 1985. Artificial marine reefs of Catalina Island: Recruitment, habitat specificity and population dynamics. Bulletin of Marine Science 37:387–395.

Zar, J. H. 1984. Biostatistical analysis, 2nd edition. Prentice-Hall, Engelwood Cliffs, New Jersey.

Social and Economic Evaluation of Artificial Aquatic Habitats

J. WALTER MILON
Department of Food and Resource Economics
University of Florida
Gainesville, Florida

I. Introduction

This chapter addresses the role of social and economic evaluation methods as they apply, or could apply, to artificial habitat development projects such as artificial reefs and fish-aggregating devices (FADs). The primary focus is on monitoring, impact assessment, and efficiency analysis, since these evaluation activities are most directly related to measuring social performance. These three methods have a long history of use in the evaluation of water and, to a somewhat lesser degree, fishery development projects around the world (e.g., Hufschmidt *et al.*, 1983; Anderson, 1986). But their application in planning and assessing artificial habitat projects has been sparse. And the different methods are often confused by fishery managers and planners who have little knowledge of social science research.

Artificial habitats may be designed for a variety of biological functions, but they primarily serve social functions by providing services or resources to different user groups. While biological evaluation is concerned with the effects of artificial habitats on aquatic organisms, social and economic evaluation seeks to understand how a project influences society and how society benefits or loses. Social and economic evaluation is based on the application

Artificial Habitats for Marine and Freshwater Fisheries
Copyright © 1991 by Academic Press, Inc.

of social science research methods designed to gather relevant and reliable evidence on the social impacts of a project. This evidence may include the impacts on economic, demographic, and social structures.

In the context of artificial habitat development, evaluation methods can be used to address questions such as: Which user groups use an artificial habitat, and what is the intensity of use? Is the artificial habitat meeting the needs of the user group it was designed to serve? Has the artificial habitat project caused the social and economic impacts it was designed to achieve? Has the expenditure of funds for artificial habitat development been an efficient expenditure? And what rules and regulations can be used to assure that the development and use of an artificial habitat are consistent with society's objectives?

These questions that attempt to identify the effects of artificial habitat development on society are becoming increasingly important as development efforts evolve from small, isolated initiatives by groups of fishermen to large-scale programs administered by government resource management agencies. Despite the recent surge in construction of new artificial habitats, there have been relatively few documented social and economic evaluations of these developments. Less than 4% of more than 400 artificial habitat research papers reviewed by Bohnsack and Sutherland (1985) addressed social and economic evaluation issues. Little has changed in the intervening period to alter the imbalance. A review of artificial habitat programs in the Southeastern and Mid-Atlantic regions of the United States, where development projects are proliferating, concluded that, "Evaluation of artificial reef programs . . . is largely nonexistent" (Murray, 1989, p. 15).

In light of the sparse use and limited knowledge of social and economic evaluation methods, this chapter provides a guide to the application of these research methods for artificial habitat projects. Examples from the artificial habitat literature are used to illustrate important concepts but the coverage is incomplete due to the lack of documented research. Most of this research is based on marine habitat projects in the United States since very few evaluation studies from other countries have been reported in the literature. This chapter is not intended to be a complete guide to social and economic evaluation methods, therefore, references to more detailed works are provided. This information can be useful to analysts concerned about the evaluation of habitat development projects, policy makers, financial supporters, and the users of artificial habitats.

The chapter is organized as follows: (1) It explains the vital linkage between habitat development objectives and monitoring, impact, and efficiency evaluation methods. Project objectives may be general or specific and they will be influenced by various rights of ownership and control of the habitat. However, the success of a project in providing services to society cannot be evaluated without a clear statement of objectives. This point is essential. (2) It describes the different approaches to monitoring and reviews some of the studies that have used these methods. (3) Next is a consideration of the different methods for impact evaluation and discussion of some important methodological problems in earlier impact evaluation studies. (4) It then explains the use of cost-effectiveness and cost-benefit analysis for economic efficiency evaluation and reviews some of the recent literature on methods that measure the benefits of artificial habitats. (5) This chapter explores the role of social and economic evaluation in habitat management and the use of management controls to restrict uses that are inconsistent with project objectives. (6) The final section presents a summary that includes critical remarks about the current state of social and economic evalaution for artificial habitat development.

II. Purposes of Social and Economic Evaluation

Ideally, social and economic evaluation should provide three types of information: (1) information that can be duplicated by different investigators, (2) information on testable hypotheses on the net outcomes of a project, and (3) information about the economic efficiency of the project (Freeman *et al.*, 1979). In addition, to be effective socioeconomic evaluation should be closely tied to the project objectives and focused on an identifiable fishery management problem.

A. Project Objectives and Evaluation

As in other areas of fishery management, the problem definition and objectives for artificial habitat efforts may be general or specific. It is important to recognize that more specific objectives provide more substantive indicators of project performance. For example, the objective stated in the United States National Artificial Reef Plan " . . . to enhance recreational and commercial fishing opportunities" (U.S. Department of Commerce, 1985, p. 5), identifies a general purpose and target user groups to benefit from artificial habitat development. This is an identifiable objective but it is not very useful as an objective for a specific project. The term "enhance" is vague and lacks a precise indicator of success or failure.

Alternatively, a project objective could be defined as "increase the exploitable biomass of species x by recreational anglers in community y by 25% using the most cost-effective artificial habitat design." This objective specifies the target user group and species, the impacted area, the desired change in the species population, and an efficiency criterion. This latter element of the objective is especially important because it encourages habitat developers to consider the cost of alternative habitat designs and provides a specific monetary criterion for evaluation.

In some cases a monetary criterion may not be important but a nonmonetary objective can still be specific. For example, the objective of an artificial habitat project may be to increase recreational diver satisfaction from diving trips by xx%. Since a user group's satisfaction can be measured across time periods or different communities (Manning, 1986), a specific nonmonetary indicator of project performance is possible. Other nonmonetary indicators of social performance might include measures of user groups' well-being, aesthetic perceptions, or other attitudinal indicators (Gregory, 1987). These nonmonetary measures of performance are valid evaluation criteria if they are specifically identified when the project objectives are defined.

In some situations the objective of artificial habitat development may relate to the distribution of benefits and the social structure in the fishery. The term social structure includes the processes of interaction between members of social groups and organizations, community norms and institutions, and community perceptions and attitudes (Leistritz and Murdock, 1981). For example, in an area where small-scale fishermen must compete with large-scale operations, the objective of an artificial habitat project may be to sustain or improve the viability of the small-scale operators. Bombace (1989) explained how such an objective was achieved by establishing artificial habitats tailored to the fishing gear and patterns of the inshore fishing group in Italy. The project was evaluated by comparing economic returns to small-scale fishermen using the artificial habitats and by assessing their willingness to cooperate in further artificial habitat development.

Artificial habitat projects also may seek to achieve multiple objectives. A habitat may be designed to increase recreational catch and to increase sport divers' satisfaction. Both objectives can be achieved through careful consideration of the specific products the project will provide for each user group and by instituting management controls to minimize the potential negative interaction between the user groups (Samples, 1989).

B. Ownership, Control, and Evaluation

The purpose of socioeconomic evaluation may depend on the pattern of ownership and control of an artificial habitat site. As with most marine and freshwater resources, artificial habitats are common property resources (Samples and Sproul, 1985). However, private groups of fishermen may be able to build artificial habitats and exclude others from using them, and the primary concern for evaluation is whether the habitats satisfy the private group's objectives. The ability to control access to the habitat may be due to direct private ownership of the aquatic system, isolation from other user groups, exclusive user rights granted by a government authority, or *de facto* tenurial relationships enforced by custom or violence (L. Sprague, personal communication). Similar institutional processes have been developed to limit access to other common property resources (e.g., National Research Council, 1986).

However, excluding "outsiders" from a habitat site is often difficult. Different user groups may exploit the site, so no one group of users has an incentive to privately finance habitat development for their own use. Therefore, most artificial habitats are local public goods, and government agencies typically have authority over the development and management. Public funding and management facilitate development, but evaluation is necessary to guide the process and assess whether public funds are used for the desired objectives.

C. Matching Evaluation with the Objectives

The type of evaluation should be closely matched with the objectives of the project and the specific geographic area impacted by it. Evaluations can be grouped into the following three general categories: monitoring, impact assessment, and efficiency analysis. Table 7.1 provides a general overview of their purpose and the primary methods used for each type of evaluation.

TABLE 7.1
An Overview of Social and Economic Evaluation Methods

Type of evaluation	Purpose	Primary methods
Monitoring	Identify habitat user groups, usage rates and determinants of use	Date collection and analysis from site observation and surveys
Impact	Measure changes in economic activity or social structures	Economic base or input-output analysis; social impact analysis
Efficiency	Determine whether project provides desired outcomes for least-cost or whether monetized value of outcome exceeds cost	Cost-effectiveness or cost-benefit analysis

1. Monitoring Evaluation

This type of evaluation determines whether a habitat is meeting the design criteria and whether the target user group is actually using the habitat. Monitoring evaluation is useful for evaluating general project objectives that are not linked to specific performance criteria. Some general objectives that might be addressed with a monitoring evalaution include the following: a) increase the number of shore-accessible recreational fishing sites in coastal community A, b) provide nearshore fishing sites for small-scale coastal fishermen in coastal bay B; or c) provide separate habitat sites for recreational divers interested in spearfishing or photography in coastal town C. This type of evaluation is conducted by gathering data on users of a project by observing the site(s) or conducting interview, mail, or telephone surveys. Data analysis can identify the factors influencing decisions to use the project.

2. Impact Evaluation

Impact evaluation is directed to more specific project objectives that seek desired changes in economic activity or social structures. This type of evaluation focuses on the changes caused by the project and tries to determine whether these changes have met the specific objectives. It is particularly important to determine whether the artificial habitat project has produced more of the desired effect than would have occurred "naturally," that is, without the site development. For example, a project objective to "increase nonresident fishing trips and economic activity in coastal community A by xx%" could be evaluated with an expenditure and economic impact analysis that measured nonresident activity both before and after a habitat development project. Similarly, an evaluation of an objective "to increase small-scale commercial fishing sales from Port A by xx%" could compare sales levels after the project is initiated with preproject sales levels. The most commonly used methods for economic impact evaluations are economic base and input-output analysis. Impacts on social structures are evaluated with social impact analysis.

3. Efficiency Analysis

This third type of evaluation is appropriate for specific objectives related to the economic performance of the project. Efficiency analyses are usually classified as either cost-effectiveness or cost-benefit types of evaluation. Cost-effectiveness analysis determines whether a project has (could) produced the desired impact at least cost. Cost-benefit analysis determines whether the monetized value of a project's benefits exceed the costs. Both cost-effectiveness and cost-benefit analyses can be used to compare the performance of several artificial habitat projects and the results can be compared to efficiency analyses of other types of enhancement projects. In addition, both types of analyses can be initiated in the planning phase of artificial development to make a preliminary evaluation of whether a project is a feasible economic investment. Thus, a project objective to increase the exploitable biomass of species y in coastal bay B by xx% using the least cost alternative could be evaluated with a cost-effectiveness analysis. Alternatively, a project objective to "provide an artificial habitat site off coastal town A that has a positive ratio of benefits to costs" could be assessed with a cost-benefit analysis.

D. Planning for Evaluation

Regardless of the types of objectives and socioeconomic evaluation methods for an artificial habitat development, the objectives and evaluation methods should be clearly identified in the planning process. Evaluations without well-articulated objectives cannot be used to determine the reasons for the success or failure of a project because the desired outcomes are unknown (Ditton and Burke, 1985). However, due to the variety of political, institutional, and environmental factors that influence planning (see Chapter 1, this volume), there is no single process to match objectives and evaluation methods. Therefore, planners should recognize the integral role of social and economic evaluation and adapt the process to address this component.

The following sections of this chapter provide a more detailed description of monitoring, impact, and efficiency evaluation methods. Relevant examples from the socioeconomic literature on artificial habitats are discussed in each section to illustrate application of the methods. While it is important to recognize that the evaluation method should be matched with the type of objective, in the interest of brevity no direct reference to types of project objectives will be made in the following sections. Hopefully, with a good understanding of social and economic evaluation methods the reader can relate different types of project objectives to the appropriate method.

III. Evaluation by Monitoring

Monitoring is the least demanding of the evaluation methods and a wide variety of data collection techniques can be used individually or in combination to obtain the necessary information about usage and performance. The primary considerations in determining the type of monitoring are the desired reliability of the data, the resources available, the research skill of the evaluator(s), and the physical characteristics of the habitat site(s). Monitoring data can be collected with methods as simple as direct observations at the site to more complex methods such as interviews with users on the site or at other locations. However, as a general principle, monitoring should not be done on a onetime basis for a short interval. Successful monitoring evaluation depends on a valid data profile that is not unduly influenced by events on one or a few days during the data collection period (Selltiz *et al.*, 1976; Finsterbusch *et al.*, 1983).

A. Methods

One form of monitoring is direct observation at the site by the evaluator. The observer can record the number and type of boats and type of activity at an artificial habitat site during different time periods. This technique may be desirable if the site is easily observable and the purpose of evaluation is to identify general usage patterns. Unfortunately, direct observation may be expensive since it requires personnel on the site for extended periods of time, and it may not be practical if the site covers a large area. Direct observation usually does not provide any demographic, social, or economic information about the user, and it does not provide information about nonusers.

A variation on the direct observation method that may be useful for large, dispersed sites is aerial photography. Observations can be scheduled for selected time periods and the photo record can be used to identify the number and types of users. This technique, however, shares many of the same disadvantages of on-site observation and may be very expensive if conducted over an extended period of time.

An alternative site-based survey that is useful for monitoring is the direct interview. Interviews may be conducted directly at a site and combined with observational data, or the interview may be conducted at user access points such as marinas, ports, or entrance ways to shore-accessed habitat sites. The direct interview can be used to collect user (and nonuser) profile data on socioeconomic characteristics, expenditures, and attitudes and provides the opportunity for the interviewer to properly identify species in the kept catch. The drawbacks of the direct interview, as with those of the direct observation method, are cost-related. It may be expensive to arrange an interview schedule for a sufficient period of time to provide representative data, especially if the sites and access points are widely dispersed. In some cases interviews can be combined with other types of surveys to obtain data on future use or other information that may be too time-consuming for an interview. Monitoring data collected by direct interviews should conform to accepted statistical procedures (e.g., Rossi *et al.*, 1983).

A common technique for gathering monitoring data on artificial habitats is the mail or telephone survey. Random or stratified samples can be selected from public record lists of potential habitat users in the study area.

These lists might include recreational and commercial boat registrations, recreational and commercial fishing licenses, and sport fishing and diving club membership. Survey instruments can be designed to elicit information about usage and catch at an artificial habitat site and other natural habitat sites. In addition, user and nonuser profile information such as socioeconomic characteristics, expenditures, and attitudes can be elicited.

The principal drawbacks of mail and telephone surveys are that respondents may not be able to provide information about the use of a specific site, the timing and duration of site use over extended periods of time, or the proper identification of species in the catch. These problems can be partly overcome by limiting the duration for recall (maximum of three to six months), by sequencing distribution of the survey over major periods of use such as fishing seasons, and by using visual aids such as maps and charts. Other problems such as inadequate sample size and survey instrument bias can be addressed by following the generally accepted procedures outlined in social survey reference works such as Dillman (1978) and Rossi *et al.* (1983).

In addition to providing information about usage of a habitat, monitoring data also can be used to identify determinants of site use. The extent of this analysis will depend on the type of usage data collected. The following subsection illustrates the use of these methods by discussing findings from selected monitoring research studies. The studies cited focus on the United States, since no monitoring studies from other countries are documented in the literature. The reader should review the original reports for detailed information about data collection methods.

B. Usage Monitoring Studies

One of the first monitoring studies of artificial habitat use was Ditton and Graefe's (1978) study of bay and offshore habitats around Galveston, Texas. They used boat registration files for a proportionally stratified mail survey of Galveston area anglers to elicit information about the recreational use of inshore tire reefs, offshore oil platforms, and offshore reefs composed of sunken ships. The latter artificial habitats were particularly important because the state of Texas had recently deployed the reefs for the " . . . goal of fisheries enhancement" (Ditton and Graefe, 1978, p.2). Unfortunately, the concept of enhancement was not clearly stated in terms of specific user groups or species. The survey results revealed that tire reefs accounted for less than 3% of inshore fishing trips, whereas ship reefs were the site choice for only 5% of offshore trips. The oil platforms were used on over 50% of the offshore trips. The authors did not evaluate the monitoring information to identify reasons for the different usage rates.

The only documented, direct observation-based monitoring study was reported by Ditton and Auyong (1984) using data collected by observers on 164 standing oil and gas platforms in the Gulf of Mexico off Louisiana over a 12 month period. Observers employed by the platform operators recorded the number of boats, the number of users in the boat, type of use, and suspected target species of anglers. The recorded observations were then matched with physical information about each platform to develop a profile of platform users in terms of the types of use, distance traveled from shore, temporal use patterns, and target species preferences. The data, however, are not necessarily representative of all platforms in the Gulf because the platform operators could decide whether and how frequently to involve their personnel in data collection. In general, the results showed that the plat-forms were used primarily by recreational anglers targeting reef and pelagic species. Most anglers used platforms within 20 miles of shore. Ditton and Auyong used the monitoring data to develop a statistical model to predict the spatial distribution of site usage based on platform distance from shore, water depth, and age of the platform. But they concluded these platform characteristics were inadequate predictors of use.

However, in a subsequent analysis of the platform observation data evaluated by Ditton and Auyong, Gordon (1987) augmented the data by cal-culating the distance from a platform to the nearest launch site and by de-termining the number of other platforms located between a launch site and a specific platform. Unfortunately, these enhancements were speculative be-cause the original observation data did not include the user's actual launch site. Gordon also clustered platform structures in close proximity (less than 300 m) into a single site. With the augmented data, Gordon's predictive model showed that site usage rates increased with proximity to the shore and the number of structures at a site but decreased with the availability of other sites; however, as in Ditton and Auyong's analysis, the water depth and age of platform were not significant determinants of use.

Samples (1986) used a mail survey of anglers (identified from a previous survey of registered boaters) to determine different groups' usage of FADs in Hawaii. The FAD project was intended to " . . . increase the fishing productivity of commercial and recreational fishermen" and to reduce their " . . . inputs of time and fuel needed to catch a given quantity of fish" (p.1). The results showed that the mixed use group (both commercial and recrea-tional fishing) had the highest usage rate of FADs (37.1 visits annually) and the recreational use group had the lowest rate (17.8 visits annually). Approxi-mately 25% of each user group indicated they had increased the number of their fishing trips since the FADs were installed. Usage rates varied across different FAD sites but the most heavily used sites were near population centers. A majority of respondents (69%) felt that they caught more fish at

FADs. Almost 50% believed that their fuel and oil costs were decreased by fishing at FADs. No predictive model was developed to identify the determinants of usage at a specific FAD site.

McGurrin and Fedler (1989) used a random sample from boat registration files for two counties in southeastern Florida to identify anglers' use of a submerged petroleum production platform. (Five sections of platform had been transported to the site from Louisiana to demonstrate the feasibility of recycling obsolete platforms as artificial habitats.) Forty-six percent of private boat anglers had used artificial habitats in the study area but only 12% had used the submerged platform site. Users of the platform habitat rated fishing quality at the site higher than fishing quality at alternative sites. The authors did not identify specific determinants of site use.

Also in southern Florida, Milon (1988a) used boat registration files for a proportionally stratified mail survey of recreational boaters in Dade County. In this study, respondents were provided with a detailed map of the offshore artificial reefs and asked to report on both fishing and diving activities at specific reef sites during both the prior six-month period and the most recent trip. The latter information included the time spent at specific sites, initial launch site, size of the user group, and total number and weight of fishes harvested. These data were combined with other information about the physical characteristics of each site in a multinomial, statistical model to predict users' decisions to fish or dive at an artificial habitat and their choice of a specific site (Milon 1989b). The results showed that approximately 28% of anglers and 14% of divers used the artificial habitats. Anglers and divers' choices of specific sites were influenced by different characteristics of the sites, but travel time from the launch site and the amount of material at the site had negative effects on site usage by both groups. The age of the site (from initial placement of material) had a positive effect on both groups. Water depth of the site had a positive influence on usage by anglers but did not have a statistically significant effect on divers. Mean catch rates at a site were inversely related to site use but other analyses (Milon, 1988b,c) showed a significant positive relationship. These conflicting results may be due to a correlation between catch and site characteristics such as age, water depth, and amount of material that were not included as explanatory variables in the latter analyses.

In general, these studies indicated that monitoring evaluations could identify overall artificial habitat usage rates and the determinants of specific site use by different user groups. Both Milon and Gordon identified an inverse relationship between distance from the launch site and site usage. However, the possible linkages between other characteristics such as the amount of site material and site catch rates may confound the task of identifying the role of other site characteristics in artificial habitat use decisions.

IV. Impact Evaluation

The purpose of impact evaluation is to determine whether a project has actually caused, or could cause, a desired change in the economy or society. Unlike monitoring evaluation which is intended only to profile participation and activity associated with a project, impact evaluation is designed to measure the changes in social conditions that occur from a project. The effects or outcomes of the project are measured in terms of observable variables defined in the project objectives. If objectives are not well defined, the evaluator may have to infer what the project was intended to accomplish and the appropriate measure of impact.

A. Basis for Methods

As discussed previously, the desired outcome from a project may be expressed in either monetary or nonmonetary terms. Regardless of the outcome measure, an impact evaluation should measure only those changes that are caused by the project and control the effects of external factors that may influence the outcome measure. One way to evaluate a project's impact is to measure the appropriate indicator variable before and after the habitat is established. For example, a project objective to reduce the distance traveled by inshore fishermen should not be evaluated by comparing the distance traveled by inshore habitat users with the distance traveled by offshore fishermen or with the distance traveled by other inshore fishermen who do not use the habitat. In the first comparison, the distance traveled by offshore fishermen is irrelevant since inshore fishermen would travel shorter distances regardless of whether the habitat exists. In the second comparison, the fishermen who selected the artificial habitat may already travel less than the nonusers so the project may only relocate or concentrate their fishing activity. The proper comparison is the distance traveled by habitat users before and after the project is established so that net change in distance traveled can be measured.

Although before and after measurement is one procedure for impact evaluation, it is not usually practical. Data collection on fishing and diving activities at specific sites is not a routine part of fishing surveys, therefore, a special effort may be necessary to establish baseline measurements before the project begins. Alternatively, it may be possible to establish a control group of potential users who are restricted from actually using the project and measure their activity concurrently with the activity of users. Data for this control group could then be compared with data for habitat users to determine the impact.

If both preproject measurements and control groups are not feasible, the only alternative is careful statistical analysis of the data to identify the influence of confounding effects on the outcome measure. This type of statistical control requires further information on appropriate factors that may influence the outcome. For example, in evaluating distance traveled by inshore fishermen after an artificial habitat is established, data on boat length and target species may be used as controls to compare the distances traveled by users and non-users. These factors may influence the choice of fishing site and hence the distance traveled. If the distance traveled by users with boats of the same length and choice of target species is lower than nonusers, the conclusion that the habitat actually reduced distance traveled is more plausible than an evaluation based solely on distance traveled by all inshore users and nonusers. This type of statistical control depends on the evaluator's prior knowledge of likely confounding factors. Appropriate statistical procedures for this type of evaluation have been applied in many areas of evaluation research and are described in standard statistics texts such as Rossi *et al.* (1983), Selltiz *et al.* (1976), and Wright (1979).

The purpose of impact evaluation is different from monitoring, yet most of the data collection techniques discussed previously are appropriate for impact evaluation. Direct observations, interviews, and mail and telephone surveys can be used depending on the situation. The primary difference is that, since impact evaluation requires a comparison, the data collection process should be carefully planned so that valid statistical comparisons can be made. This means that the samples are randomly selected and the same measurement variables are collected. Planning is necessary whether the comparison is based on before and after measurements or on a cross-section analysis using statistical controls. *Ad hoc* judgments by an evaluator cannot substitute for rigorous impact evaluation.

Although several measures of social conditions can be used for impact evaluation (Vanderpool, 1987), the remainder of this section will focus on economic impact assessment. Change in economic conditions within a community is frequently discussed as an objective for artificial habitat development, but efforts to measure the actual economic impacts of artificial habitat projects have been very limited.

B. Economic Impact Evaluation—
Two Approaches

Economic impact analysis focuses on the changes in sales, income, and employment resulting from a project. It is the appropriate evaluation method when a project is intended to change the economy of a coastal community or region. The total economic impact of a project is composed of the following three types of economic impact: (1) The direct impact of a project is the initial change in local spending or final demand for goods and services; (2) Indirect impacts are the changes in output and employment to produce the goods and services required by businesses who receive the direct impact; (3) The induced impacts are the changes in demand for goods and services due to the change in local employees' income brought about by the direct and indirect impacts. The indirect and induced impacts of projects are usually referred to as secondary impacts because they only occur as a result of direct impacts. Economic impact analysis measures the total impact of a project on sales, income, and employment by summing the direct and secondary impacts.

The economic impact of a project is usually estimated with multipliers. A multiplier expresses the relationship between the direct impact of a project and the resulting changes in sales, income (wages and earnings), and employment within the impacted area. These are called the sales or output multiplier, the income multiplier, and the employment multiplier. Other multipliers can also be developed for special evaluation purposes (Stevens and Lahr, 1988). Each multiplier is a useful measure of economic change, but the choice of a specific type of multiplier for an impact analysis should be linked to whether the project is intended to increase output, income, or employment. Most development efforts seek to improve community income and the number of jobs, therefore, the income and employment multipliers are most useful.

A number of methods used to compute multipliers are discussed in greater detail in reference works such as Leistritz and Murdock (1981), Hewings (1985) and Propst and Gavrilis (1987). The two most commonly used approaches are the economic base and input–output methods. Each embodies a different view of the development process that can influence the outcome of an impact evaluation.

1. Base Theory

The base theory method divides a local economy into the following two sectors: basic and nonbasic. The basic sector is composed of industries that sell the majority of their products outside the region and might include manufacturing, agriculture, fisheries, and tourism. The nonbasic sectors produce goods for local consumption and might include services and utilities. Several different methods can be used to classify an industry as basic or nonbasic (Richardson, 1985).

In base theory the basic sector is the source of all growth in the local economy, so all economic impacts must be due to changes in basic sector sales, income, and employment. If r is defined as the ratio of basic income (employment) to nonbasic income (employment):

$$r = \frac{I_b}{I_{nb}}$$

then it can be shown (Hewings, 1985) that total income (employment), I_T, can be expressed as:

$$I_T = \left[\frac{(1 + r)}{r}\right] I_b$$

and the term in the brackets is the income (employment) multiplier. Thus, if total income is 10 million and income in the basic sector is 6 million, then r is 1.5 (6/4), and the income multiplier is 1.67. A direct change in basic sector income of 100,000 will cause an indirect change in nonbasic sector income of 67,000, and thus a total income change in the community of 167,000.

The base theory approach to impact analysis requires that an artificial habitat project must produce sales, income, and employment from outside the community for the project to cause an economic impact within the community. This means that exports of fish products or tourism from outside the region must change. Otherwise, sales within the community simply redirect the existing economic activity and have no secondary impact. In addition, the multiplier effect would be the same for any increase in basic activity regardless of whether the change resulted from fish product exports or tourists who use the habitat. Both of these restrictions can be addressed using the input–output method.

2. Input–Output

An input–output (I/O) model disaggregates the basic and nonbasic sectors into many large industry groups. In this framework, output from one industry becomes input to another and the series of interindustry transactions is driven by final household demand and exports. An I/O model for a local economy can be used to compute sales, income, and employment multipliers for each sector. [For further details on the computation of I/O multipliers, see Hewings (1985) and Miller and Blair (1985)]. Thus, with input–output multipliers it is possible to account for the impact of a project on each sector in the local economy. Also, secondary impacts of a project do not depend solely on changes in exports since it is possible to measure the impacts of changes in spending within the local economy.

Several factors determine the magnitude of a project's economic impact. Industries that are more dependent on other local industries will usually have higher multiplier effects than industries that rely on industries outside the local economy. This is because secondary impacts of the project are diminished by leakages outside the local area. Multiplier effects also will usually vary directly with the size of the local economy. Small communities usually have relatively little interdependence between industries and rely heavily on imports from other communities. On the other hand, large communities most often have greater interdependence and thus multipliers are larger. The exact influence of these factors will vary by industry and the composition of the local economy. Thus, in contrast to the implications of a base theory multiplier, secondary impacts can vary by sector so that equal increases in exports and tourism caused by a habitat development project may have different impacts on the community. Also, boundaries defined for the local economy can influence the magnitude of the impacts. These differences should be considered when project objectives related to economic development are established.

C. Economic Impact Studies

There are only two documented studies of economic impacts of artificial habitats, both completed for the U.S. recreational fishery on the Atlantic coast. The first combined direct observation with a mail survey of anglers in Murrells Inlet, South Carolina (Buchanan, 1973). Since baseline data prior to development of the habitat were not available, the mail survey used a "with or without" approach by asking tourist anglers whether they would return to Murrells Inlet if the Paradise Reef artificial habitat was not present. Those who indicated they would not return were classified as the group whose fishing effort was influenced by the habitat. Results of the study showed that nonresident anglers who would not return took fewer trips than reef users who would return and anglers who did not use the reef (Table 7.2). The group who would not return accounted for only 8.7% of the direct expenditures by tourist anglers. Buchanan did not compute the secondary impacts of these direct impacts on the Murrells Inlet economy. In addition, the study did not identify whether the anglers who would not return would simply relocate their trips to an adjacent coastal community. If the tourist anglers relocated, the effect of the habitat was a redistribution of fishing effort with no net economic impact across all the coastal communities.

Liao and Cupka (1979) used a mail survey of registered boaters in South Carolina to identify the economic impact of the state's artificial reef program. This survey did not distinguish resident from tourist anglers. Baseline data were not available so anglers were divided into two groups—reef users and nonusers. The authors then used output multipliers from a state level input–output model to estimate the secondary output impacts of the expenditures

TABLE 7.2
Nonresident Anglers' Economic Activity, Per Capita, in Murrells Inlet,
South Carolina, without Paradise Reef[a]

	Reef Anglers		Nonreef Anglers
	Not Return	Return	
Average number of days fishing per year	14.0	71.8	29.5
Average expenses per day (1972 U.S. $)	$21.44	$8.55	$14.74
Percent of total nonresident expenditures	8.7 %	80.0 %	11.3 %

[a] Adapted from Buchanan, 1973.

by reef users. Twenty-two percent of total economic activity due to sport fishing in the state was attributed to the artificial habitats. Unfortunately, this study was conceptually flawed because the researchers did not attempt to determine, as Buchanan did, whether the anglers would continue to fish even if the artificial reefs were not present. Since fishing effort and expenditures may have been the same with or without the reefs, no clear measure of economic impact was established.

V. Efficiency Evaluation

Although the purpose of impact evaluation is to measure the effects of a project on economic activity and social structure in a community, it does not include a specific criterion to determine whether the project is beneficial or harmful. As such, impact analysis is "value-free" in the sense that no value judgments are imposed by the analytical methods. By contrast, efficiency evaluations such as cost-effectiveness and cost-benefit analysis impose a standard to evaluate whether a project is good or bad. The standard imposed by cost-effectiveness analysis differs from the standard imposed by cost-benefit analysis. Cost-effectiveness seeks the alternative that produces the desired outcome at the least cost, whereas cost-benefit analysis determines whether the monetary value of the outcome justifies the project cost.

A. Methods

The choice of an evaluation standard depends on the objectives of the project and the characteristics of benefits it produces. As a general rule, cost-effectiveness analysis is most useful when a specific tangible outcome is desired that cannot be easily measured in monetary terms. For example, the use of alternative habitat designs to provide habitat for an endangered species could be evaluated by comparing the costs per fish protected. The least-cost design per fish protected would be preferred. Cost-benefit analysis is more appropriate when project benefits can be monetized and are spread across several user groups. For example, alternative artificial habitat designs to increase the exploitable biomass of a species for recreational and commercial harvesting can be compared with a cost-benefit analysis. The design with the highest ratio of benefits to costs would be preferred.

One element of efficiency evaluation that is common to both cost-effectiveness and cost-benefit analyses is proper accounting for the social costs of a project. This appears to be a deceptively simple element of efficiency evaluation because many indirect project costs are often overlooked, resource costs are not always fully reflected in their prices, and costs incurred over the life of the project are often neglected.

Costs of an artificial habitat project include the full range of resources used, from initial design and planning stages to the final stage including the cost of removing the habitat from the water, if appropriate. The major categories of costs listed in Table 7.3 are representative of the costs typically incurred in an artificial habitat project. Many discussions of artificial habitat costs have focused only on material and transportation costs (e.g., Prince and Maughan, 1978; Shomura and Matsumoto, 1982) and neglected the personnel costs to plan, design, administer, and implement the project. In addition, maintenance costs over the life of the project for elements such as marker buoys, repositioning, or other activities after the initial deployment often are overlooked.

Evaluation—whether for monitoring, impact assessment, or efficiency analysis—is also a necessary part of project development and should be reflected in project costs. And the external costs caused by a project such as material removal from a beach after a storm or damage to vessels are part of project costs. While these external costs may not be known until the project is implemented, recognition that these are legitimate project costs encourages artificial habitat planners to consider the full costs of particular designs. Some external costs may be avoided by proper planning and siting.

Costs such as building rent or dockage fees may be easy to measure because their price reflects the cost of using the resource. But other resources may be imperfectly priced or not priced at all so that the true opportunity cost of the resource is not counted. The opportunity cost is the economic return a resource would bring in an alternative use. For example, an abandoned ship may be available for an artificial reef project. Since the

TABLE 7.3
List of Items to Include as Artificial
Habitat Project Costs

Personnel
+ Planning and design
+ Administration
+ Boat operator(s)
+ Laborers

Facilities
+ Administrative offices
+ Warehousing and assembly areas
+ Docking

Materials
+ Acquisition
+ Preparation

Transportation
+ To assembly and preparation area
+ To deployment site

Maintenance
+ Marker bouys
+ Repairs

Insurance
+ Liability coverage

Evaluation
+ Monitoring
+ Impact
+ Efficiency

External costs

Dismantling and removal costs

ship is not owned, there are no out-of-pocket costs for acquisition. However, it is important to recognize that the ship could be sold for scrap and the proceeds used to buy a prefabricated habitat structure. Since the ship has an alternative use, the opportunity cost is the scrap value and this cost should be included in the total costs. Similarly, the opportunity cost of an obsolete offshore oil and gas platform that might be used for an artificial reef is equal to the onshore salvage value minus dismantling and transportation costs. From a social accounting perspective, this opportunity cost is the value of the resource that should be counted as an acquisition cost in an efficiency analysis. Examples of other ways to measure opportunity costs are discussed in greater detail in texts on efficiency evaluation such as Thompson (1980).

It is important also to count costs over the full time period of the project. Costs can be incurred for maintenance, replacement, evaluation, and removal at the end of the project. Depending on the length of time the project will operate, this part of the total costs may be substantial. A further complication arises from the fact that a cost incurred in later years of a project is not as costly as an initial cost because of the effects of discounting. Discounting is a means of accounting for the effects of time on the value of money by expressing future costs (or benefits) as their "present value." For example, if the known useful life of a project was three years, the removal costs were $10,000, and the cost of money was 10%, the present value of these future costs would be $7513. Further aspects of the timing of project costs are discussed in the following section on cost-benefit analysis.

1. Cost-Effectiveness Analysis

Cost-effectiveness analysis compares the costs of alternative means of achieving the same goals. It is usually used in one of two settings: as a preproject evaluation to determine the least-cost choice from available alternatives to accomplish the same objective, or as a postproject evaluation to compare the costs per unit of outcome to a known alternative or a standard defined in the project objectives. In both cases the analysis focuses on a cost-effectiveness ratio defined as:

$$\frac{\text{net costs}}{\text{net units of outcome}}$$

For example, suppose we want to compare the cost-effectiveness of a new prefabricated artificial reef design to a scrap reef design. The expected net increase in harvested biomass for the prefabricated design is 10,000 kg per year and the total discounted cost is $100,000. Monitoring data for the scrap reef revealed that harvested biomass was 8000 kg per year and the total cost of the scrap reef was $40,000. Thus, the two cost-effectiveness ratios are as follows:

$$\text{Prefab design:} \quad \frac{\$100,000}{10,000 \text{ kg}} = \$10 \text{ per kg}$$

$$\text{Scrap design:} \quad \frac{\$40,000}{8,000 \text{ kg}} = \$5 \text{ per kg}$$

show that the scrap design is more cost-effective since the cost per unit of catch is lower than the alternative.

A different comparison might have been made if the scrap reef cost and biomass data were not available. For this comparison the expected costs per unit could be compared to a cost standard given in the project objectives. If the cost per unit was lower than the standard, the prefab design would be cost-effective. Thus, the outcome of an evaluation to determine whether a project is cost-effective depends on the alternative(s) considered. A single cost-effectiveness ratio has little value for evaluation purposes unless it can be compared to an alternative or a desired standard.

2. Cost-Benefit Analysis

Cost-benefit evaluation depends on the simple principle that project benefits in excess of costs leads to an improvement in social well-being. This principle has been applied for many years in water and fishery resource development projects around the world (e.g., Hufschmidt *et al.*, 1983; Bishop *et al.*, 1990). The primary advantage of cost-benefit analysis is that it can be used to express a broad array of benefits received and costs incurred by different social groups in a single monetary measure. The limitation is that it is sometimes difficult to express the benefits of a project in monetary terms. However, many of these difficulties can be overcome with appropriate methods to measure economic benefits, and recent progress in measuring the benefits of artificial habitats indicates that these projects are no exception.

The benefits of artificial habitat projects include tangible outcomes such as change in commercial fishery harvests due to a habitat, and intangible outcomes such as the increased enjoyment a sport diver experiences from photographing a habitat. Tangible effects are valued by measuring the net amount of money gained from the outcome, for example, an increase in commercial fishery profits or net income. Intangible effects are valued by measuring the beneficiaries' net willingness to pay for the outcome. In some cases intangible effects may be too nebulous to measure, such as endangered species protection or research. In these situations, cost-effectiveness analysis may be a more appropriate evaluation tool.

It is important to distinguish cost-benefit analysis from economic impact analysis because the two types of evaluation are commonly confused. As discussed previously, economic impact analysis assesses direct and secondary changes in sales, income, and employment caused by a project. Increases in sales, income, and employment are usually considered to be beneficial for the local community. However, increases in one community may be matched by offsetting declines in economic activity elsewhere (Talhelm, 1985). This result might occur in the case of two adjacent coastal towns where an artificial habitat project in one of the towns attracts tourists to the project but at the expense of tourism in the neighboring town (see the earlier discussion of Buchanan, 1973).

In addition, impact evaluation does not consider the costs of implementing the project. Cost-benefit analysis considers both sides of the ledger and counts only the direct impacts on income that result from the project.

Other categories of what might be considered beneficial impacts such as sales and employment growth are not counted because these effects are caused by the change in income.

Focusing on the direct income changes in cost-benefit analysis avoids the problem of double-counting benefits and provides a consistent basis to compare benefits and costs (Edwards, 1990). Proper use of either impact analysis or cost-benefit analysis depends on a clear understanding of the purpose for the evaluation.

The benefits from an artificial habitat generally will accrue to one or more user groups that can be described collectively as commercial fishermen, sport anglers and sport divers. Benefits to commercial fishermen will accrue from two possible outcomes as follows: (1) the harvested biomass per unit effort is higher at the artificial habitat site, while harvesting costs per unit effort are the same as at natural habitat sites, and (2) harvested biomass per unit effort is constant but costs are lower due to fuel, labor, or gear savings. These outcomes (discussed more fully in Chapter 5) lead to changes in fishermen's net income or profits, or, more formally, the "producers' surplus." Producers' surplus can be defined as the difference between total revenues and total costs where total revenue is the market price per unit of harvested biomass times the total units of biomass harvested (Milon, 1989a). Market price of the biomass reflects society's value or willingness to pay for the products of the habitat. Where biomass has no market price and is used for subsistence, its value can be estimated using various "shadow pricing" methods (Hufschmidt et al., 1983; Thompson 1980). The net benefits for commercial fishermen can thus be defined as:

$$\sum_t \sum_n (PS_{tn})$$

where PS is the change in producers' surplus due to the habitat, $t = 1, \ldots, T$ denotes time periods and T is the useful life of the habitat, and $n = 1, \ldots, N$ denotes the total number of commercial boats using the habitat. Thus, commercial fishing benefits from an artificial habitat depend upon the net change in producers' surplus for all boats using the habitat over the project's useful life.

Benefits to sport anglers and divers also accrue in one or a combination of two ways: (1) harvested biomass or enjoyment from a trip to an artificial habitat site is higher than a trip to a natural habitat site and trip costs are the same, or (2) harvested biomass or enjoyment is the same, but trip costs are lower. The important difference between commercial and recreational user benefits from an artificial habitat is that the latter do not depend on changes in catch or on the market price of the catch. Instead, these benefits are measured by the sport users' willingness to pay, or "consumers' surplus," for the habitat site. This surplus is the amount sport users are willing to pay for the habitat over and above the actual expenditures incurred in using the habitat. This surplus can be measured for both resident and tourist sport users because each group can directly benefit from the project. This is an important distinction from economic impact evaluation that attributes beneficial impacts solely to new tourist spending. Note also that the consumer surplus measure of benefits does not include trip expenditures, since these expenditures could be incurred elsewhere.

Using the consumers' surplus measure of sport benefits, total sport users' benefits from an artificial habitat site can be expressed as:

$$\sum_t \sum_s (CS_{ts})$$

where $t = 1, \ldots, T$ again denotes the useful life of the project, $s = 1, \ldots, S$ denotes the total number of sport users, and CS is the change in consumers' surplus or net willingness to pay for the site (Milon, 1989a). These sport user benefits can be estimated with two different approaches: the travel cost method and contingent valuation. The travel cost method is based on actual site choice decisions, whereas contingent valuation uses hypothetical decision scenarios to elicit willingness to pay. Walsh (1986) and Mitchell and Carson (1989) provide introductory discussions on the use of these approaches for a broad array of resource and environmental valuation problems. Bockstael *et al.* (1985) discuss the applicability and limitations for estimating sport users' benefits from artificial habitats.

Commercial and sport user benefits from an artificial habitat site can be aggregated in a summary indicator to evaluate the efficiency of the project. Using the formulas for commercial and sport user benefits described above, the cost-benefit ratio for a project is given by the following equation:

$$\frac{\sum_{t=1}^{T} \left(\left(\sum_{n=1}^{N} PS_{tn} \right) + \left(\sum_{s=1}^{S} CS_{ts} \right) \right) \Big/ (1 + r)^t}{C_{t=0} + \sum_{t=1}^{T} C_t \Big/ (1 + r)^t}$$

where $C_t = 0$ is the first-year cost to establish the habitat, $C_t = 1, \ldots, T$ are the annual maintenance costs over the life of the habitat, and $(1 + r)^t$ is the discount rate based on the borrowing costs, r, for project funding. The ratio indicates that a project with discounted benefits in excess of discounted costs is efficient. This approach to efficiency evaluation is useful because it can integrate the benefits to different user groups over the full, useful life of the project. The cost-benefit ratio can be used for pre- or postproject evaluation, although it may be difficult to use in preproject settings where the biological

effects are highly uncertain and potential user groups are not familiar with
the effects of the technology. Cost-benefit evaluation is especially useful for
comparing artificial habitats with other fishery enhancement programs such
as hatcheries and natural habitat protection (see Bishop *et al.*, 1990).

B. Efficiency Evaluation Studies

Despite the importance of efficiency evaluation in other water resource
development areas, there have been relatively few studies for artificial habitat projects. Fiegenbaum *et al.* (1989) reported a cost-effectiveness analysis
of tire and concrete habitat designs used in the Chesapeake Bay of the U.S.
mid-Atlantic Coast. They compared the (undiscounted) cost per unit of catch
for the alternative designs over a 50-year potential useful life. Although the
tire and concrete designs had similar catch rates and the concrete designs
cost more initially, the concrete designs were more cost-effective because
they were less likely to deteriorate.

In some studies the concepts of efficiency evaluation were improperly
applied. Bell *et al.* (1989) reported a preliminary cost-effectiveness analysis
of manufactured concrete, plastic, and steel artificial reef designs. No harvestable catch data were available for the comparative analysis but the authors did compare the cost of the manufactured designs to Liao and Cupka's
(1979) estimated economic impacts from the South Carolina artificial reef
program. Bell *et al.* (1989) mistakenly refer to these estimated impacts as
the " . . . total economic value to the state" (p. 826) and fail to recognize that
economic impact estimates should not be used for efficiency evaluations (see
earlier discussion). Nor do they question the dubious basis for Liao and Cupka's estimate, also discussed previously. Thus, their conclusion, "It is certainly reasonable, based on these economic estimates, for the state to utilize
manufactured reef units to replace or supplement the use of scrap materials
in future artificial reef construction efforts" (p.826), is not justified by the
evidence.

A similar problem occurs in the efficiency evaluation guidelines for preproject siting decisions proposed by Chang (1986). Chang defines the benefits of an artificial habitat to include (1) the (commercial) value of fish harvested; (2) a measure of the consumers' surplus for the habitat; and (3) the
expenditures by tourists using the reefs. This approach incorrectly mixes
efficiency evaluation with impact evaluation because tourist benefits are correctly estimated with (2). Part (3) expenditures are the costs of angling, not
benefits from the project. Including these costs would result in overinflated
benefit estimates. In addition, Chang's proposed approach would lead to the
same error as the Liao and Cupka (1979) study cited earlier by not identifying whether the level of fishing effort would actually change because of the

artificial habitat. The guidelines also err in the computation of total project costs. Chang proposes that subsidies by government or charity organizations to defray the costs of habitat development should be deducted from the total costs. This would be correct if the habitat development was a private venture, but since the purpose is social evaluation the subsidies are only transfers that do not change the true costs of the project to society. Most cost-benefit analysis practitioners recommend that transfers should not be deducted from total project costs for social efficiency evaluation (e.g., Freeman *et al.*, 1979).

Another improper use of efficiency evaluation methods occurs in artificial habitat studies in Japan. Reported studies of Japanese artificial habitat technology (e.g., Mottet, 1982; Ohshima, 1982; Sato, 1985) have compared total revenues from harvested biomass per unit of habitat to the total costs per unit. Since the economic benefits of the habitat should be measured by the change in producers' surplus (profits) and profits are only a small fraction of total revenues, the total revenues overestimate economic benefits from the habitats. Thus, even though Japan has the largest government financed artificial habitat development program in the world (Yamane, 1989), there have been no studies presented at international conferences or translated from the Japanese language demonstrating the economic efficacy of this program.

A few research projects in the United States have attempted to measure recreational benefits of artificial habitats using the travel cost and contingent valuation approaches. Results of these studies are summarized in Table 7.4. The estimated (mean) annual willingness to pay by users of artificial habitats ranged from U.S. $1.80 to U.S. $491. Nonusers' (mean) willingness to pay varied from U.S. $2.93 to U.S. $226. Several observations about these estimates are important. First, the studies show that the travel cost and contingent valuation approaches can provide similar estimates of willingness to pay for the same location.

The studies by Bockstael *et al.* (1986) and Milon (1988c, 1989c) used both methods and produced valuation estimates that were comparable for the study location, but the estimates were significantly different across the two locations. However, the range of estimates reported in Milon (1988c, 1989c) also show that there is likely to be considerable variation even for the same location due to variations in the application of the methods by the researcher.

In addition, the Bockstael *et al.* (1986) and Milon (1989c) studies showed that nonusers were willing to pay for artificial habitats even though they had not used them in the past. Although the nonusers' willingness to pay was lower than that of users, the estimates indicate that nonusers perceive some benefits from habitat development due to expected future use or stock enhancement. Unfortunately, the studies cover such a broad range of habitats

TABLE 7.4
Valuation Estimates for Uses of Artificial Habitats from Selected Studies in the United States (Reported Mean Values)

Study and location	Type of artificial habitat	Users' activities	Valuation method	Results
Roberts and Thompson (1983)—Louisiana	Oil drilling platforms (standing)	Sport diving	Contingent valuation	WTP[a] for right to use offshore platforms Users—$163
Bockstael et al. (1986)—South Carolina	Offshore sites: tires and derelict vessels	Sport fishing	Travel cost	WTP to prevent loss of sites Users—$491 Nonusers—$196
			Contingent valuation	WTP to maintain sites Users—$328 Nonusers—$226
Samples (1986)—Hawaii	Offshore floating bouys	Sport and commercial fishing	Contingent valuation	WTP to maintain buoys Users—$89
Milon (1988c)—Florida	Offshore sites: derelict vessels and steel debris	Sport fishing	Travel cost	WTP for new habitat site Users—$1.80 to $38.59[b]
Milon (1989c)—Florida	Offshore sites: derelict vessels and steel debris	Sport fishing and diving	Contingent valuation	WTP for new habitat site Users —$15.71 to $27.38[b] Nonusers—$2.93 to $17.30[b]

[a] WTP—Annual willingness to pay.
[b] Range of mean values due to variation in application of valuation method.

(oil platforms, FADs, and artificial reefs) that it is not possible to compare the estimates to determine whether the differences across studies can be attributed to specific features of the habitats.

Although these studies provided theoretically correct measures of the benefits of artificial habitats, the only study to report a complete cost-benefit evaluation was Samples (1986). He used a mail survey to elicit sport and commercial anglers' net willingness to pay for the Hawaiian FAD network. The estimated annual benefits (U.S. $184,906 in 1985) slightly exceeded average annual costs over the first five years of the project (U.S. $182,000). Samples noted that project administration and planning costs accounted for 38% of total costs and caused the average cost per FAD to be significantly higher than cost estimates for other FAD systems by Shomura and Matsumoto (1982) who did not include these costs. Samples also noted that a related study by Sproul (1984) estimated that commercial tuna fishermen increased their producers' surplus by using FADs (U.S. $70,236 in 1983) so that the project was an efficient use of public funds.

Finally, Bombace (1989) reported that commercial users of artificial reefs in the Mediterranean Sea had net proceeds 2.5 times larger than other fishermen and that annual benefits were 2.3 times larger than annual costs. However, no details were provided about the methods used to estimate benefits or the expenses that were accounted for in the total costs.

VI. Evaluation and Habitat Management

Although the purpose of socioeconomic evaluation is to determine whether a project is satisfying the planning objectives, an evaluation may also identify problems that interfere with performance. Like other types of aquatic resources, artificial habitats may require some degree of management control to protect the biological resource and regulate conflicts within and between user groups. These conflicts occur because the habitats are created by public agencies in public waters without effective restrictions on access by user groups. Social and economic evaluation can identify the nature of these conflicts and management action can help to resolve them.

A. Stock and Congestion Conflict

Samples (1989) classifies user conflicts as stock or congestion effects depending on whether the conflict results from reductions in fishery stock at an artificial habitat site or too many users in a limited space. Stock effects may occur from overharvesting of all species at a site or from selective overharvesting of particular species or sizes. Congestion effects occur when one user's activity interferes with another's. Conflict may result from incompatible uses (e.g., sport diving and commercial fishing), incompatible fishing gears, or simply too many users at a site. Stock and congestion effects are not mutually exclusive but social and economic evaluations can help to identify the source of conflict. For example, a monitoring evaluation at a habitat site could identify incompatible or unintended uses. An efficiency evaluation could be used to identify the economic costs of conflicts and differences in the values each user group derives from the habitat. This economic information can be useful in determining whether benefits from management outweigh the costs of establishing and enforcing controls.

Samples (1989) outlines different types of habitat management alternatives to deal with stock and congestion effects (Table 7.5). Conflicts over stock effects can be reduced by limiting harvests by one or all user groups. These harvest reductions can be attained by setting catch limits, by limiting fishing gear types, or by taxing harvests. Samples (1989, p. 849) suggests that the most practical alternative may be gear restrictions since this type of control is commonly used to restrict overharvesting.

Controls to reduce congestion effects are less concerned about harvest and more concerned about interaction between users. This interaction can be limited by requiring users to have licenses and setting a maximum number of licenses. Also, restrictions on fishing gear can reduce conflicts by preventing gear types that are not compatible with the objectives of the artificial habitat project or program. By limiting the time of access of each user group, they can continue to use the habitat without interaction between groups. The full extent of this approach could be achieved by spatially segregating users with separate habitat sites for each user group. Samples (1989) concludes that no single management control will be optimal for all situations, therefore, the choice of one or more controls must be based on an evaluation to determine the nature of the conflict and the effectiveness of a control strategy. In some cases it may not be socially beneficial to control these conflicts.

In practice, several types of management approaches have emerged to avoid or limit user conflicts. In some countries conflicts are avoided because a single user group has exclusive access rights to a habitat. For example, Japanese cooperatives are granted exclusive commercial fishing rights to regions of the coastline, thereby prohibiting other user groups from harvesting from artificial habitat sites (Polovina and Sakai, 1989). In other countries, gear and catch restrictions have been used to limit access. The South Atlantic Fishery Management Council in the United States has designated some artificial habitat sites as special management zones where commercial harvesting gear is prohibited. The harvest of certain reef fish species within these zones has also been prohibited (McGurrin, 1989; National Marine

TABLE 7.5
Habitat Management Controls to Reduce User Conflict
(adapted from Samples, 1989)

Stock effects	Congestion effects
Harvest limits by user group (number and size)	Licensing
	Gear restrictions
Gear restrictions	Temporal segregation
Types of harvests	Spatial segregation

Fisheries Service, 1990). In Malaysia, all fishing is prohibited within one-half mile of an artificial habitat site (Hung, 1988). Further socioeconomic evaluation research is necessary to determine whether these and other management controls successfully limit user group conflicts.

VII. Summary and Discussion

The purpose of this chapter was to provide an overview of social and economic evaluation methods and a review of recent applications for artificial habitat developments. The key points can be summarized as follows: (1) To determine the social effectiveness of a habitat development, evaluation must be closely tied to the objectives of the project. (2) The type of evaluation will depend on whether project objectives are general or specific, and also many unique features of the project such as ownership and control of the habitat, site characteristics, and constraints on pre- and postproject data collection for the evaluation will shape the evaluation.

Monitoring evaluations are useful for assessing the congruence of a development with objectives related to the design criteria and the target user groups for the development. These evaluations can be used to identify the characteristics of the user groups, activities they pursue, users' attitudes and opinions about the habitat, and the characteristics of the habitat that attract users. This type of information provides valuable feedback to project administrators and may lead to reassessment of the initial objectives or a redesign of the project.

Impact evaluations allow project administrators to gauge the extent of social change caused by a project. They are intended to determine whether the project produced the desired effects and had the outcome(s) defined in the project objectives. Impacts may be defined in a number of social and economic dimensions including social structure, indicators of user satisfaction, and indicators of economic activity in the impacted region. Economic impacts on sales, income, and employment can be assessed with economic base or input-output multipliers but the two methods provide significantly different levels of detail about impacts and can lead to different measures of impacts.

Efficiency evaluations can be used to determine whether the project has been an effective expenditure of funds and to provide a basis for comparing project performance to other fishery enhancement technologies. Cost-effectiveness analysis does not attempt to measure the project outcomes in monetary terms but measures the outcomes produced per monetary unit of cost. This is a limited form of efficiency analysis that is useful for comparing two or more alternatives. Cost-benefit analysis provides a monetary comparison of the benefits to the initial and recurring costs over the project life time. Both cost-effectiveness and cost-benefit analysis can be undertaken as pre- or postproject evaluations to guide and inform the project design and administration process.

Social and economic evaluations also can be helpful in designing management controls for the habitat. Due to the open-access nature of artificial habitats in most areas, regulations or other social institutions may be necessary to assure the habitat provides the desired outcomes for target user groups. Information from evaluations can be used to help resolve user conflicts caused by open-access.

Current research helps to provide some insights about the role of social evaluation methods in artificial habitat developments, but it is apparent that more evaluations are necessary and the quality of these evaluations could be improved. Many prior efforts to provide data for evaluations have not followed generally accepted sampling procedures (e.g., Ditton and Auyong, 1984; McGurrin and Fedler, 1989), or else sufficient information about sampling procedures was not provided to assess the quality of the data. As a result, the information reported has limited value and may misrepresent the actual performance of an artificial habitat. In some cases, methodological errors may have provided project managers with misleading information about project performance (e.g., Liao and Cupka, 1979). In part, these problems are common to most new areas of research. However, the problems affecting the artificial habitat development process appear to be especially onerous.

Many socioeconomic evaluations of artificial habitat projects have been based on the evaluator's interpretation of the project's objective because the objectives were not clearly defined in a written plan. All of the existing economic evaluation research has focused on short-term effects with no empirical evaluation of long-term effects. None of the existing studies involved follow-up evaluations that are necessary for sound scientific analysis. Aside from Samples' (1986) evaluation of FADs in Hawaii, there are no complete, well-documented studies of the economic impacts or economic efficiency of artificial habitats, despite the frequent assertion of beneficial effects by program administrators. In part, these shortcomings are attributable to a research agenda that has been dominated by biological concerns with limited attention to the social dimensions of habitat development.

The most serious problem, however, is the apparent lack of understanding of the use of socioeconomic evaluations by artificial habitat program administrators. Murray's (1989) review of programs in the Southeastern and Mid-Atlantic states of the United States suggests that administrative interest in socioeconomic evaluation is primarily to justify new funding initiatives (p. 15). There is very little in the published literature to contradict the impression that the same attitude prevails in other areas. It is essential that program administrators and others involved in artificial habitat development recognize the limitations of this perspective and actively seek to develop a research agenda for socioeconomic evaluation in the habitat planning and development process (Milon, 1989a).

The promise of artificial habitat technologies is that they can provide a means to expand fishery stocks and control the harvests of these stocks. Wise and judicious use of these technologies depends on sound biological and social research to document the true impacts and to provide a basis for comparison with other fishery enhancement techniques. The overview of socioeconomic evaluation methods in this chapter provides a basis for understanding the role of these research methods and a guide to further reading on their proper implementation.

References

Anderson, L. G. 1986. The economics of fisheries management, 2nd edition. Johns Hopkins University Press, Baltimore, Maryland.

Bell, M., C. J. Moore, and S. W. Murphey. 1989. Utilization of manufactured reef structures in South Carolina's marine artificial reef program. Bulletin of Marine Science 44:818–830.

Bishop, R. C., S. R. Milliman, K. J. Boyle, and B. L. Johnson. 1990. Benefit-cost analysis of fishery rehabilitation projects: A Great Lakes case study. Ocean and Shoreline Management 13:253–274.

Bockstael, N., A. Graefe, and I. Strand. 1985. Economic analysis of artificial reefs: An assessment of issues and methods, Technical Report 5. Artificial Reef Development Center, Sport Fishing Institute, Washington, D.C.

Bockstael, N., A. Graefe, I. Strand, and L. Caldwell. 1986. Economic analysis of artificial reefs: A pilot study of selected valuation methodologies, Technical Report 6. Artificial Reef Development Center, Sport Fishing Institute, Washington, D.C.

Bohnsack, J. A., and D. L. Sutherland. 1985. Artificial reef research: A review with recommendations for future priorities. Bulletin of Marine Science 37:11–39.

Bombace, G. 1989. Artificial reefs in the Mediterranean Sea. Bulletin of Marine Science 44:1023-1032.

Buchanan, C. C. 1973. Effects of an artificial habitat on the marine sport fishery and economy of Murrells Inlet, South Carolina. Marine Fisheries Review 35(9):15–22.

Chang, S. 1986. Siting plans for the establishment of artificial reefs in the Gulf of Mexico: An economic analysis. Pp. 433–538 in J. I. Jones, editor. A plan for siting artificial reefs in the northern Gulf of Mexico, Publication MASGP-86-021. Mississippi-Alabama Sea Grant Consortium, Ocean Springs, Mississippi.

Dillman, D. A. 1978. Mail and telephone surveys—the total design method. Wiley, New York.

Ditton, R. B. and J. Auyong. 1984. Fishing offshore platforms—central Gulf of Mexico, OCS Monograph MMS84–0006. U.S. Department of the Interior, Minerals Management Service, Metairie, Louisiana.

Ditton, R. B., and L. B. Burke. 1985. Artificial reef development for recreational fishing: A planning guide. Sport Fishing Institute, Washington, D.C.

Ditton, R. B., and A. Graefe. 1978. Recreational fishing use of artificial reefs on the Texas Gulf Coast. Texas Agricultural Experiment Station, Texas A&M University, College Station.

Edwards, S. F. 1990. Allocating fish stocks between commercial and recreational fisheries: An economics primer. U.S. National Marine Fisheries Service, Report NMFS 94, Woods Hole, Massachusetts.

Fiegenbaum, D., M. Bushing, J. Woodward, and A. Friedlander. 1989. Artificial reefs in Chesapeake Bay and nearby coastal waters. Bulletin of Marine Science 44:734–742.

Finsterbusch, K., L. G. Llewellyn, and C. P. Wolfe, editors. 1983. Social impact assessment methods. Sage Publications, Beverly Hills, California.

Freeman, H. E., P. H. Rossi, and S. R. Wright. 1979. Evaluating social projects in developing countries. Development Centre of the Organization for Economic Co-operation and Development, Paris.

Gordon, W. R., Jr. 1987. Predicting recreational fishing use of offshore petroleum platforms in the central Gulf of Mexico. Ph.D. Dissertation, Texas A & M University, College Station.

Gregory, R. 1987. Nonmonetary measures of nonmarket fishery resource benefits. Transactions of the American Fisheries Society 116:374–380.

Hewings, G. J. D. 1985. Regional input-output analysis. Sage Publications, Beverly Hills, California.

Hufschmidt, M. M., D. E. James, A. D. Meister, B. T. Bower, and J. A. Dixon. 1983. Environment, natural systems and development: An economic valuation guide. Johns Hopkins University Press, Baltimore, Maryland.

Hung, E. W. F. 1988. Artificial reef development and management in Malaysia. In Report of the Workshop on Artificial Reef Development and Management, ASEAN/SF/88/GEN/8. Penang, Malaysia.

Leistritz, L. F., and S. H. Murdock. 1981. The socioeconomic impact of resource development: Methods for assessment. Westview Press, Boulder, Colorado.

Liao, D. S., and D. M. Cupka. 1979. Economic impacts and fishing success of offshore sport fishing over artificial reefs and natural habitats in South Carolina, Technical Report 38. South Carolina Marine Resources Center, Charleston.

Manning, R. E. 1986. Studies in outdoor recreation—a review and synthesis of the social science literature in outdoor recreation. Oregon State University Press, Corvallis.

McGurrin, J. 1989. An assessment of Atlantic artificial reef development. Fisheries 14(4):19–25.

McGurrin, J. M., and A. J. Fedler. 1989. An evaluation of rigs-to-reefs in fisheries development. Pp. 131–137 in V. C. Reggio, compiler. Petroleum structures as artificial reefs: A compendium. OCS Study/89–0021. U.S. Department of the Interior, Minerals Management Service, New Orleans, Louisiana.

Miller, R. E., and P. D. Blair. 1985. Input-output analysis: Foundations and extensions. Prentice-Hall, Englewood Cliffs, New Jersey.

Milon, J. W. 1988a. The economic benefits of artificial reefs: An analysis of the Dade County, Florida reef system, Report 90. Florida Sea Grant College, Gainesville.

Milon, J. W. 1988b. A nested demand shares model of artificial habitat choice by sport anglers. Marine Resource Economics 5:191–213.

Milon, J. W. 1988c. Travel cost methods for estimating the recreational use benefits of artificial marine habitat. Southern Journal of Agricultural Economics 20:87–101.

Milon, J. W. 1989a. Economic evaluation of artificial habitat for fisheries: Progress and challenges. Bulletin of Marine Science 44:831–843.

Milon, J. W. 1989b. Artificial marine habitat characteristics and participation behavior by sport anglers and divers. Bulletin of Marine Science 44:853–862.

Milon, J. W. 1989c. Contingent valuation experiments for strategic behavior. Journal of Environmental Economics and Management 17:293–308.

Mitchell, R. C., and R. Carson. 1989. Using surveys to value public goods: The contingent valuation method. Resources for the Future, Washington, D.C.

Mottet, M. G. 1982. Enhancement of the marine environment for fisheries and aquaculture in Japan, Technical Report 69. Washington State Department of Fisheries, Olympia.

Murray, J. D. 1989. A policy and management assessment of Southeast and Mid-Atlantic artificial reef programs, Publication UNC-SG-WP-3. Raleigh. University of North Carolina Sea Grant.

National Marine Fisheries Service. 1990. The potential of marine fishery reserves for reef fish management in the U.S. southern Atlantic, NOAA Technical Memorandum NMFS-SEFC-261. Southeast Fisheries Center, Miami, Florida.

National Research Council. 1986. Proceedings of the conference on common property resource management. National Academy Press, Washington, D.C.

Ohshima, Y. 1982. Report from the consolidated reef study society. Pp. 93–98 in S. Vik, editor. Japanese artificial reef technology, Technical Report 604. Aquabio, Inc., Belleair Bluffs, Florida.

Polovina, J. J., and I. Sakai. 1989. Impacts of artificial reefs on fishery production in Shimamaki, Japan. Bulletin of Marine Science 44:997–1003.

Prince, E. D., and O. E. Maughan. 1978. Freshwater artificial reefs: Biology and economics. Fisheries 3:5–9.

Propst, D. B., and D. G. Gavrilis. 1987. Role of economic impact assessment procedures in recreational fisheries management. Transactions of the American Fisheries Society 116:450–460.

Richardson, H. W. 1985. Input-output and economic base multipliers: Looking backward and forward. Journal of Regional Science 25(4):607–61.

Roberts, K. and M. E. Thompson. 1983. Petroleum production structures: economic resources for Louisiana sport divers. Louisiana Sea Grant College Program, LSU-TL-83-002, Baton Rouge. 39 pp.

Rossi, P. H., J. D. Wright, and A. B. Anderson, editors. 1983. Handbook of survey research. Academic Press, New York.

Samples, K. C. 1986. A socioeconomic appraisal of fish aggregation devices in Hawaii, Marine Economics Report 33. University of Hawaii Sea Grant College Program, Honolulu.

Samples, K. C. 1989. Assessing recreational and commercial conflicts over artificial fishery habitat use: Theory and use. Bulletin of Marine Science 44:844–852.

Samples, K. C., and J. T. Sproul. 1985. Fish aggregating devices and open-access commercial fisheries: A theoretical inquiry. Bulletin of Marine Science 37:305–317.

Sato, O. 1985. Scientific rationales for fishing reef design. Bulletin of Marine Science 37:329–335.

Selltiz, C., L. S. Wrightsman, and S. W. Cook. 1976. Research methods in social relations. Holt, Rinehart and Winston, New York.

Shomura, R., and W. Matsumoto. 1982. Structured flotsam as aggregating devices, NOAA Technical Memorandum NOAA-TM-NMFS-SWFC-22. U.S. National Marine Fisheries Service, Southwest Fisheries Center, Honolulu, Hawaii.

Sproul, J. 1984. Estimating net benefits received by Hawaii's pole and line tuna vessel owners as a result of utilizing a network of fish aggregation devices: 1980–1983. M.S. Thesis, University of Hawaii, Honolulu.

Stevens, B. H., and M. L. Lahr. 1988. Regional economic multipliers: Definition, measurement, and application. Economic Development Quarterly 2:88–96.

Talhelm, D. R. 1985. The economic impact of artificial reefs on Great Lakes sport fisheries. Pp. 537–543 in F.M. D'Itri, editor. Artificial reefs: Marine and freshwater applications. Lewis Publishers, Inc., Chelsea, Michigan.

Thompson, M. S. 1980. Benefit-cost analysis for program evaluation. Sage Publications, Beverly Hills, California.

U.S. Department of Commerce. 1985. National artificial reef plan, NOAA Technical Memorandum NMFS OF-6. U.S. National Marine Fisheries Service, Washington, D.C.

Vanderpool, C. K. 1987. Social impact assessment and fisheries. Transactions of the American Fisheries Society 116:479–485.

Walsh, R. G. 1986. Recreation economic decisions: Comparing benefits and costs. Venture Publishing, State College, Pennsylvania.

Wright, S. R. 1979. Quantitative methods and statistics: A guide to social research. Sage Publications, London.

Yamane, T. 1989. Status and future plans of artificial reef projects in Japan. Bulletin of Marine Science 44:1038-1040.

Index

Abalone
 artificial habitats of the world and, 49
 artificial reef ecology and, 96
Abiotic factors
 artificial reef ecology and, 62, 93
 environmental assessment and, 223
  data needs, 190–191, 193–196
  methods, 221–223
  problems, 186
  purpose, 183
Abundance
 artificial habitats of the world and, 31
 artificial reef ecology and, 62
  assemblage structure, 65–66
  design, 77–78, 80
  environment, 71
  population dynamics, 93, 97
 environmental assessment and
  data needs, 189, 191–192, 195
  methods, 202–203, 205, 208, 210–211
  visual census methods, 213, 216
 fisheries applications and, 168
Acadja, artificial habitats of the world and, 38
Acanthurus spp., 208
Adaptation, artificial reef ecology and, 61, 75
Africa, brush parks and, 38
Aggregation, *see* Fish aggregation; Fish aggregation devices
Algae
 artificial reef ecology and
  assemblage structure, 64, 66
  environment, 72–75
  population dynamics, 93, 95
 engineering of manufactured habitats and, 142
 environmental assessment and, 192, 221
Algeria, 37

Ambloplites rupestris, 75
American Petroleum Institute (API), engineering of manufactured habitats and, 112–113
American Samoa, 35, 159
Archosargus probatocephalus, 49
Artificial habitat practices, 1–2
 diversity, 2–7
 issues, 21–22
  enhanced design, 25–27
  function, 22–24
  planning, 24–25
 planning, 13–17
 purpose, 13
 research, 17–21
 worldwide utilization, 8–13
Artificial habitats of the world, 31–32
 case studies
  Japan, 44–49, 57
  United States, 44, 49–57
 philosophical bases, 43–44
 scope, 32–33
  bibliographies, 41–43
  international conferences, 39–41
  local efforts, 38–39
  national programs, 33–36
  regional approaches, 36–38
Artificial Reef Development Center, 53
Artificial reefs
 artificial habitat practices and, 6, 19–21, 23–25
 artificial habitats of the world and, 35–37, 39
  bibliographies, 41–43
  case studies, 45–46, 50–54, 56–57
 engineering of manufactured habitats and, 147, 149

271